AGRICULTURE ISSUES AND POLICIES

THE CONSERVATION RESERVE PROGRAM

ISSUES AND CONSIDERATIONS

AGRICULTURE ISSUES AND POLICIES

Additional books in this series can be found on Nova's website under the Series tab.

Additional e-books in this series can be found on Nova's website under the e-book tab.

AGRICULTURE ISSUES AND POLICIES

THE CONSERVATION RESERVE PROGRAM

ISSUES AND CONSIDERATIONS

PRESTON L. CRUZ
EDITOR

nova publishers
New York

Additional color graphics may be available in the e-book version of this book.

Library of Congress Cataloging-in-Publication Data

ISBN: 978-1-63482-282-4

Published by Nova Science Publishers, Inc. † New York

CONTENTS

PREFACE

This book discusses the issues and considerations of the CRP.

Chapter 1 - The Conservation Reserve Program (CRP) provides payments to agricultural producers to take highly erodible and environmentally sensitive land out of production and install resource conserving practices for 10 or more years. CRP was first authorized in the Food Security Act of 1985 (P.L. 99-198, 1985 farm bill) and is administered by the U.S. Department of Agriculture's (USDA's) Farm Service Agency (FSA) with technical support from other USDA agencies. Participants offer land for enrollment through two types of sign-up: general and continuous. General sign-ups are competitive and only open during select times. Continuous sign-ups are not competitive, always open for enrollment, and offer additional financial incentives to those who qualify. Continuous sign-ups are targeted to specific environmental and resource concerns and operate through a number of initiatives. The largest and most well-known is the Conservation Reserve Enhancement Program (CREP), which partners with states to address agricultural-related environmental concerns in specific geographic regions. While the majority of current acres enrolled were under general sign-ups (19.7 million acres), an increasing number are enrolled under continuous sign-ups (5.8 million acres).

The Agricultural Act of 2014 (2014 farm bill, P.L. 113-79) reauthorized CRP and reduced the enrollment cap from 32 million acres to 24 million acres by FY2018. The 2014 farm bill made several changes centered on permitted activities. Emergency harvesting, grazing, and other use of forage are permitted, in some cases, without a reduction in rental rate, as well as livestock grazing for a beginning farmer or rancher. Other approved activities, such as annual or routine grazing, may continue to require a reduction in rental rate. The Grassland Reserve Program (GRP) was repealed in the 2014 farm bill.

Grassland contracts, similar to what was repealed under GRP, are now eligible under CRP. The 2014 farm bill also allows CRP participants the opportunity to terminate their contract early if the land has been enrolled longer than five years and it does not contain environmentally sensitive practices.

A number of factors have impacted CRP enrollment recently, mainly high commodity crop prices. These strong prices have encouraged farmers to put CRP acres, even marginal acres, back into production. This pressure could potentially reduce the number of CRP acres offered for reenrollment once they have expired or cause existing current CRP participants to seek an early release from their CRP contract. Some participants also have cited a potentially low CRP rental rate compared to the market rental rate as a reason for decreased enrollment interest. Despite these factors, enrollment has increased under continuous sign-ups and demand for the program, in general, still exceeds the current enrollment level.

CRP has contributed to a number of environmental benefits including reduced soil erosion, improved water quality through wetlands and field buffers, reduced fertilizer use, and increased wildlife habitat. The recent expiration of a number of acres from the program, and a reduced reenrollment, has some concerned that a number of the environmental benefits gained under CRP could be lost or reduced if land is returned to production.

Chapter 2 - This report considers how increased commodity prices might influence enrollment in and benefits from the Conservation Reserve Program (CRP) using two complementary models: a *likely-to-bid* model that uses National Resources Inventory data to simulate offers to the general signup portion of the CRP and an *opt-out* model that simulates retention of current CRP contracts. Under several higher crop price scenarios, including one that incorporates 15 billion gallons of crop-based biofuels production, maintaining the CRP as currently configured will lead to significant expenditure increases. If constraints are placed on increasing rental rates, it might be possible to meet enrollment goals with moderate increases in CRP rental rates—but this will mean accepting lower average Environmental Benefits Index scores as landowners with profitable but environmentally sensitive lands choose not to enroll.

Chapter 3 - The U.S. Department of Agriculture spends over $5 billion per year on conservation programs, mostly on voluntary programs that give financial assistance to farmers and landowners to provide environmental services (such as implementing nutrient management programs or planting native grasses). Since most programs cannot fund all interested parties, program managers must use some mechanism to select applicants. One option is to elicit offers through an auction. This report addresses the use of auctions in conservation programs. It considers how information in the hands of Government officials and rural landowners affects the auction's performance, and how auction design can reduce Government expenditures or encourage landowners to provide greater environmental services. Results of laboratory experiments are discussed, highlighting shortcomings of common features of conservation program auctions (such as limits on the rent landowners may request), as well as how alternative auction designs can improve performance.

In: The Conservation Reserve Program
Editor: Preston L. Cruz

ISBN: 978-1-63482-282-4

Chapter 1

CONSERVATION RESERVE PROGRAM (CRP): STATUS AND ISSUES*

Megan Stubbs

SUMMARY

The Conservation Reserve Program (CRP) provides payments to agricultural producers to take highly erodible and environmentally sensitive land out of production and install resource conserving practices for 10 or more years. CRP was first authorized in the Food Security Act of 1985 (P.L. 99-198, 1985 farm bill) and is administered by the U.S. Department of Agriculture's (USDA's) Farm Service Agency (FSA) with technical support from other USDA agencies. Participants offer land for enrollment through two types of sign-up: general and continuous. General sign-ups are competitive and only open during select times. Continuous sign-ups are not competitive, always open for enrollment, and offer additional financial incentives to those who qualify. Continuous sign-ups are targeted to specific environmental and resource concerns and operate through a number of initiatives. The largest and most well-known is the Conservation Reserve Enhancement Program (CREP), which partners with states to address agricultural-related environmental concerns in specific geographic regions. While the majority of current acres enrolled

* This is an edited, reformatted and augmented version of a Congressional Research Service publication R42783, prepared for Members and Committees of Congress, dated August 29, 2014.

were under general sign-ups (19.7 million acres), an increasing number are enrolled under continuous sign-ups (5.8 million acres).

The Agricultural Act of 2014 (2014 farm bill, P.L. 113-79) reauthorized CRP and reduced the enrollment cap from 32 million acres to 24 million acres by FY2018. The 2014 farm bill made several changes centered on permitted activities. Emergency harvesting, grazing, and other use of forage are permitted, in some cases, without a reduction in rental rate, as well as livestock grazing for a beginning farmer or rancher. Other approved activities, such as annual or routine grazing, may continue to require a reduction in rental rate. The Grassland Reserve Program (GRP) was repealed in the 2014 farm bill. Grassland contracts, similar to what was repealed under GRP, are now eligible under CRP. The 2014 farm bill also allows CRP participants the opportunity to terminate their contract early if the land has been enrolled longer than five years and it does not contain environmentally sensitive practices.

A number of factors have impacted CRP enrollment recently, mainly high commodity crop prices. These strong prices have encouraged farmers to put CRP acres, even marginal acres, back into production. This pressure could potentially reduce the number of CRP acres offered for reenrollment once they have expired or cause existing current CRP participants to seek an early release from their CRP contract. Some participants also have cited a potentially low CRP rental rate compared to the market rental rate as a reason for decreased enrollment interest. Despite these factors, enrollment has increased under continuous sign-ups and demand for the program, in general, still exceeds the current enrollment level.

CRP has contributed to a number of environmental benefits including reduced soil erosion, improved water quality through wetlands and field buffers, reduced fertilizer use, and increased wildlife habitat. The recent expiration of a number of acres from the program, and a reduced reenrollment, has some concerned that a number of the environmental benefits gained under CRP could be lost or reduced if land is returned to production.

The Conservation Reserve Program (CRP) is the largest federal, private-land retirement program in the United States. The program provides financial compensation for landowners to voluntarily remove land from agricultural production for an extended period (typically 10 to 15 years) for the benefit of soil and water quality improvement and wildlife habitat.

The program was first authorized in the Food Security Act of 1985 (1985 farm bill, P.L. 99-198), initially as both a supply management tool for removing land from agricultural production, thus lowering commodity supply and potentially raising prices, and for providing environmental benefits.

Currently, close to 25.6 million acres are enrolled in the program with total funding of approximately $2 billion annually.

Acres enrolled in CRP have shown a number of positive environmental benefits including reduced soil erosion; water quality improvements through vegetative cover, buffer strips, and reduced fertilizer application; and wildlife population improvement from increased habitat. While a number of natural resource improvements are attributed to the program, the program contains a number of controversial elements as well, including the economic and environmental effect of permitted activities, such as haying and grazing on CRP acres and the reduction of enrolled acres due to high crop prices and farm bill reauthorization. Program and funding authority for CRP was reauthorized and extended through FY2018 by the Agricultural Act of 2014 (2014 farm bill, P.L. 113-79).

HOW CRP WORKS

The program is administered by the Farm Service Agency (FSA) at the U.S. Department of Agriculture (USDA), with technical support from the Natural Resources Conservation Service (NRCS) and other USDA agencies.

Enrollment

Enrollment is limited to no more than 27.5 million acres at any given time in FY2014.[1] There are two main types of enrollment into CRP: general sign-up and continuous sign-up. Several continuous sign-up "initiatives" focus enrollment on specific resource concerns or conservation practices.

General Sign-Up

CRP is a competitive program, in which landowners offer eligible land for enrollment into the program. A general sign-up is a specific period of time during which FSA accepts these offers. Offers are ranked according to an Environmental Benefits Index (EBI, see text box) to determine the relative environmental benefits for the land offered.

For each general sign-up, FSA collects data on each of the EBI factors and ranks all eligible offers across the country. After the sign-up ends, USDA determines an EBI threshold. Acceptance for enrollment into CRP is extended to offers that scored above the EBI threshold. This threshold varies by sign-up depending on the offers received. Producers generally try to maximize EBI points and increase the likelihood that their offer will be accepted for enrollment.

As of July 2014, 19.7 million acres were enrolled in CRP under general sign-up contracts, or 77% of total CRP acres. This includes 262,417 contracts on 177,983 farms.[3] During the most recent general sign-up (#45), USDA accepted 1.68 million acres offered for enrollment starting October 1, 2013.

On June 4, 2014, USDA announced that contract holders of eligible FY2014 expiring contracts will qualify for a one-year extension. A general sign-up is not scheduled for FY2014.[4]

Continuous Sign-Up

Continuous sign-up is designed to enroll the most environmentally desirable land into CRP through specific conservation practices or resource needs. Unlike the general sign-up process, land offered under continuous sign-up may be enrolled at any time and is not subject to competitive bidding. If offers meet certain eligibility requirements then they are automatically accepted. Contracts are effective the first day of the month following the month of approval and typically include additional financial incentives.

Continuous sign-up includes a number of initiatives that target acres with specific resource concerns or support additional conservation practices. These are described in *Appendix A*. As of July 2014, 5.75 million acres were enrolled in CRP under continuous sign-up, or 23% of total CRP acres. This includes 1.6 million enrolled through the two statutorily created sub-programs—the Conservation Reserve Enhancement Program (CREP, 1.3 million acres) and Farmable Wetland Program (FWP, 339,673 acres).[5] The remaining 4.2 million acres were enrolled in other continuous sign-up initiatives.

Environmental Benefit Index (EBI)

Following the 1990 farm bill (P.L. 101-624), CRP was required to consider the environmental benefits of the land offered for enrollment. The formulation of the EBI has changed over time, including becoming more transparent to participants. Generally, the EBI is a standardized way to compare different land types with different resource needs across the country. The EBI is designed to compare the benefits that offered land can provide.

Presently, FSA collects data for each of the EBI factors for the land offered.

These factors are weighted and scored based on the land's potential to generate the desired environmental benefits. Some factors are made up of sub-factors (listed in parentheses). The most recent general sign-up (#45) included the following factors and weights:[2]

- Wildlife Factor evaluates the expected wildlife benefits of the offer (wildlife habitat cover benefits, wildlife enhancement, and wildlife priority zones): 10-100 points
- Water Quality Benefits Factor evaluates the potential impact that the offer may have on both ground and surface water quality (location, ground water quality, and surface water quality): 0-100 points
- Erosion Factor evaluates the potential for the land to erode from wind or water and is measured using an erodibility index: 0-100 points
- Enduring Benefits Factor evaluates the likelihood for certain practices to remain in place beyond the CRP contract period (weighted average for all practices): 0-50 points
- Air Quality Benefits Factor evaluates the air quality improvements made by reduced particulate matter and increased carbon sequestration (wind erosion impacts, wind erosion soils list, air quality zones, and carbon sequestered): 3-45 points
- Cost of environmental benefits per dollar expended (cost of the offer and how much the offer is below the maximum payment rate): 0-25 points.

Eligibility

Producer/Landowner

To be eligible for CRP enrollment, a producer must be an owner, operator, or tenant of the land for at least 12 months prior to the close of the CRP sign-up period, and show control of the land for the duration of the contract. The land may be eligible if owned for less than 12 months and if (1) the land was acquired due to the previous owner's death; (2) the ownership change occurred due to foreclosure where the owner exercised a timely right of redemption in accordance with state law; or (3) adequate assurances are made that the new owner did not acquire the land for the purpose of placing it in CRP.[6]

Land

For land to be eligible for CRP, USDA may consider the following land types for enrollment:[7]

- highly erodible cropland that (1) if untreated could substantially reduce the land's future agricultural production capability or (2) cannot be farmed in accordance with a conservation plan;[8] and has a cropping history or was considered to be planted for four of the six years preceding February 7, 2014 (except for land previously enrolled in CRP);
- marginal pasture land devoted to appropriate vegetation for water quality purposes;
- grasslands that (1) contain forbs or shrubland on which grazing is the predominant use; (2) are located in an area historically dominated by grasslands; and (3) could provide habitat for ecologically significant animal and plant populations if restored or retained in its current condition.
- cropland that is otherwise ineligible, if it is determined that (1) if permitted to remain in agricultural production, it would contribute to the degradation of soil, water, or air quality; (2) the land is a newly created, permanent grass sod waterway, or a contour grass sod strip; (3) the land will be devoted to newly established living snow fences, permanent wildlife habitat, windbreaks, shelterbelts, or filterstrips or riparian buffers devoted to trees or shrubs; (4) the land poses an off-farm environmental threat; or (5) enrollment of the land would facilitate a net savings in groundwater or surface water resources; or
- certain land enrolled as a riparian buffer or for similar water quality purposes.

Payments

In exchange for enrollment into CRP, participants receive payments from USDA. These payments offset the cost of temporarily retiring the land from production and implementing resource-conserving and wildlife-promoting practices. A number of payment types under CRP are highlighted in *Table 1.*

The authorizing statute establishes the maximum number of acres that can be enrolled in the program at any one time.[9] The program is authorized to spend such sums as necessary to enroll up to the maximum level of allowable acres. This funding is mandatory (i.e., not subject to annual appropriations) and is provided through the borrowing authority of the USDA's Commodity Credit Corporation (CCC). In total, the average annual federal cost for CRP is close to $2 billion. The majority of this cost is annual rental payments, which average $63.65 per acre, but can vary greatly by location.[10]

Table 1. CRP Payments

Payment Type	Description	Limit	Sign-up Type
Rental Payment	Annual payment to participants. Based on soil productivity for each county and the average dryland case rental rate.	$50,000 annually for any person or legal entity	general and continuous sign-up
Cost-share Payment	Payment for a percentage of installing or establishing an eligible practice.	No more than 50% of the actual or average cost of establishing the practice.	general and continuous sign-up
Maintenance Incentive Payment	Reimburses participants for the average annual cost of certain practice maintenance.	$5 per acre per year	certain continuous sign-up practices
One-time Sign-up Incentive Payment (SIP)	One-time incentive payment made to participants that enroll certain practices.	$10 per acre per year enrolled (not to exceed 10 years)	certain continuous sign-up practices.
One-time Practice Incentive Payment (PIP)	One-time incentive payment for eligible installation costs for certain practices	40% of the eligible cost of practice installation	certain continuous sign-up practices.
Other Financial Incentive	Additional incentives, as part of annual rental payments, for windbreaks, grass waterways, filter strips, and riparian buffers	Up to 20% of the annual rental payment	certain continuous sign-up practices.

Sources: 16 U.S.C. 3834, 7 C.F.R. 1410.40-1410.42, and USDA, FSA, Conservation Reserve Program Continuous Sign-Up, Fact Sheet, July 2010, http://www.fsa.usda.gov/Internet/FSA_File/crp_contsignup_072610.pdf.

Practices

Producers have a number of conservation practices to consider for installation on their land when enrolling in CRP. The selection of practices is part of the voluntary enrollment process and is determined by the landowner, with assistance from USDA, while developing a CRP offer. Once an offer is accepted for enrollment, the participant must develop a conservation plan of operation, which serves as a guide for which practices will be used, where, and for how long. Once the plan is approved and the contract signed by the participant, the land is considered enrolled in CRP. Certain continuous sign-up initiatives require specific conservation practices for enrollment. The most widely applied conservation practices are described in *Table 2.*

Table 2. Top Five Conservation Practices Installed on CRP Acres (current as of July 2014)

Practice Code	Practice Description	Acres Enrolled	Leading States
CP2	Establishment of permanent native grasses	6,753,649	Texas, Colorado, Kansas
CP10	Already established vegetative cover	4,289,400	Texas, Colorado, Montana
	(grasses and legumes)		
CP1	Establishment of permanent introduced grasses and legumes	3,370,780	Missouri, Montana, Washington
CP4D	Permanent wildlife habitat	2,126,750	Colorado, North Dakota, Kansas
CP25	Rare and declining habitat	1,708,108	Kansas, Nebraska, Minnesota

Source: USDA, FSA, Conservation Practices Installed on CRP (acres), Cumulative, July 2014.

Note: Based on total acres enrolled in practices for all sign-up types.

Current Issues

Farm Bill Reauthorization

The 2014 farm bill reauthorized CRP and reduced the enrollment cap from 32 million acres to 24 million acres by FY2018. While CRP enrollment has fluctuated since its creation in the 1985 farm bill, recent enrollment has declined from its peak in FY2007 of 36.8 million acres to 25.6 million acres in FY2013. Reduced enrollment is thought to be a product of high commodity prices, low rental rates, and declining interest in retiring land from production. Further reduction in the farm bill was viewed as inevitable given the fiscal challenges. Conservation and wildlife groups, however, remain concerned that reduced enrollment will impact critical species habitat and soil and water quality. The 2014 farm bill enrollment reduction created an estimated savings of $3.3 billion over 10 years.

A number of programmatic changes centered around permitted activities. Emergency harvesting, grazing, and other use of forage are permitted, in some cases, without a reduction in rental rate, as well as livestock grazing for a beginning farmer or rancher. Other approved activities, such as annual or routine grazing, may continue to require a reduction in rental rate (discussed further in the "Harvesting and Grazing" section below). The Grassland Reserve Program (GRP) was repealed in the 2014 farm bill. Grassland contracts, similar to what was repealed under GRP, are terminate their contract early if the land has been enrolled longer than five years and it does not contain environmentally sensitive practices. A detailed analysis of the programmatic changes may be found in *Appendix B*.

Harvesting and Grazing

Harvesting and grazing became a major concern beginning in the summer of 2012. A prolonged drought and flooding in parts of the Midwest had livestock owners requesting access to land enrolled in CRP for forage harvesting and grazing. While USDA did allow emergency harvesting and grazing under CRP, annual rental rates were reduced due to the statutory requirement that payments be reduced commensurate with the economic value of the authorized activity. Historically, this reduction ranged between 10% and 25% of the annual rental payment.

Following amendments made in the 2014 farm bill, harvesting and grazing are still permitted on CRP land under certain conditions. The amendments expand these opportunities by reducing or eliminating payment penalties and incorporating elements of the repealed Grassland Reserve Program (GRP) into grassland contracts. In some cases, environmentally sensitive land is ineligible for harvesting and grazing and most have restrictions during primary nesting season (*Figure 1*). Now rate reductions for permitted activities are as follows:

- Forage in response to drought, flooding, or other emergency—no reduction in rental rates.
- Authorized activities consistent with soil conservation, water quality, and wildlife habitat; managed harvesting; and commercial use (including biomass harvesting)[11]—not less than a 25% reduction in annual rental rates for acres covered by the activity. To occur at least every five years but not more than once every three years.
- Annual prescribed grazing for invasive species control—not less than a 25% reduction in annual rental rate, subject to nesting season restrictions.
- Routine grazing—not less than a 25% reduction in annual rental rate, subject to nesting season restrictions, vegetation management requirements and stocking rates, and limited to not more than once every two years (accounting for regional differences).
- Wind turbine installation—not less than a 25% reduction in annual rental rate, subject to nesting restrictions and limitations on the number and location or the turbines.
- Seasonal use of vegetative buffer practices—no reduction in rental rates assuming the permitted use does not damage the permanent vegetation.
- Livestock grazing by a beginning farmer or rancher—no reduction in rental rates, providing the grazing is consistent with soil conservation, water quality, and wildlife habitat; nesting season; control of invasive species; or routine grazing.

The 2014 farm bill repealed GRP and incorporated elements of the program into CRP. GRP included two enrollment types—contracts and easements.[12] GRP contracts limited future development and cropping uses of the land and retained the right to conduct common grazing practices and operations related to the production of forage and seeding.[13] A number of similar activities associated with grasslands are now permitted under CRP. These include common grazing and maintenance practices; haying, mowing, or harvesting for seed production (subject to nesting season); fire presuppression, rehabilitation, and breaks; and fencing and livestock watering. Grassland enrollment is limited to no more than 2 million acres between FY2014 and FY2018 as part of the overall program.

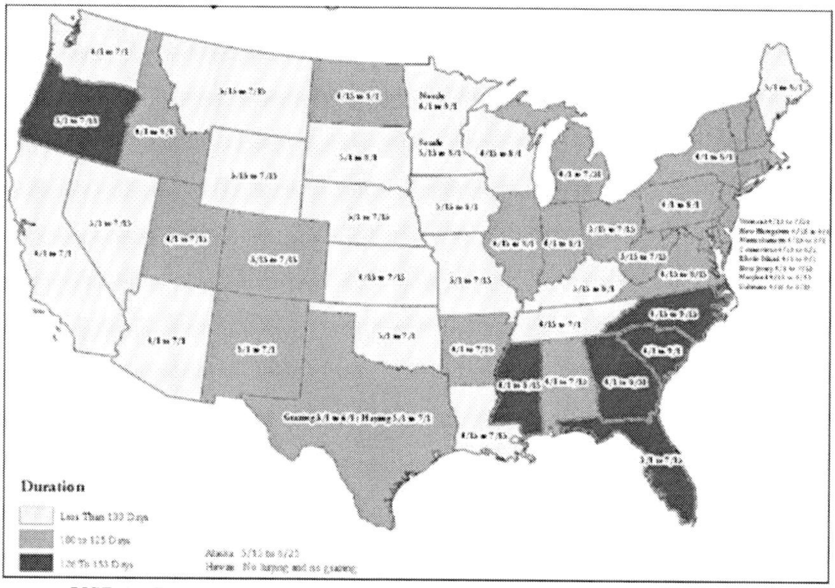

Source: USDA, FSA, DAFP, PECD, CAB, June 10, 2013, Primary Nesting Season Dates and Duration, http://www.fsa.usda.gov/Internet/FSA_ File/mapnesting 062013.pdf.

Figure 1. Primary Nesting Season Dates and Duration: Limitation on CRP Harvesting and Grazing Activities.

Recent droughts have fueled questions about the reduction of CRP acres. It is unclear what level of relief to livestock is achieved through the emergency harvesting and grazing during drought or if there is any long-term impact on wildlife habitat. Other questions remain, including, if fewer acres are enrolled in CRP for conserving uses, or if enrollment were limited to more sensitive land that would not support harvesting and grazing, what impact would this have on livestock times of prolonged drought? Is the role of CRP to provide drought relief or is that beyond the scope of the program? What are the positive and negative effects of harvesting and grazing, whether managed or in the event of emergency, that might impact wildlife, plant quality, and erosion control?

Enrollment

The nature of CRP enrollment has changed since the program's inception in the 1985 farm bill. Program priorities have shifted, total acres enrolled have fluctuated (see *Figure 2* and *Figure 3),* farming technologies have advanced, and producer preferences have changed. Many of these changes are cited as the reason for further reducing the level of CRP acres enrolled in the 2014 farm bill.

(acres by fiscal year)

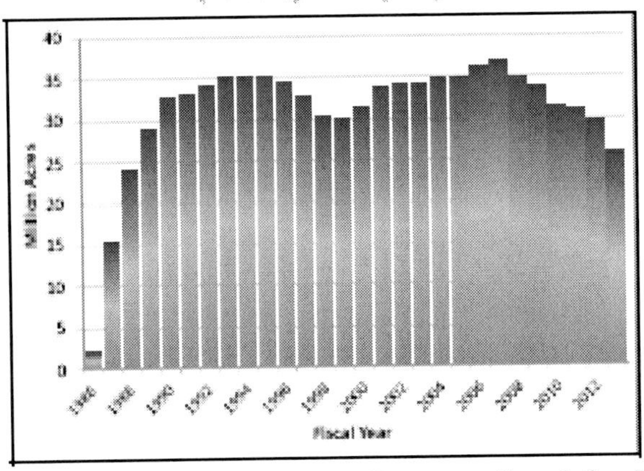

Sources: USDA, FSA, Conservation Reserve Program—Cumulative Enrollment by Year (Acres), http://www.fsa.usda.gov/Internet/FSA_File/historystate8612.xls and CRP October 2013 Monthly Summary.

Figure 2. Cumulative CRP Enrollment.

(acres enrolled as of November 2013)

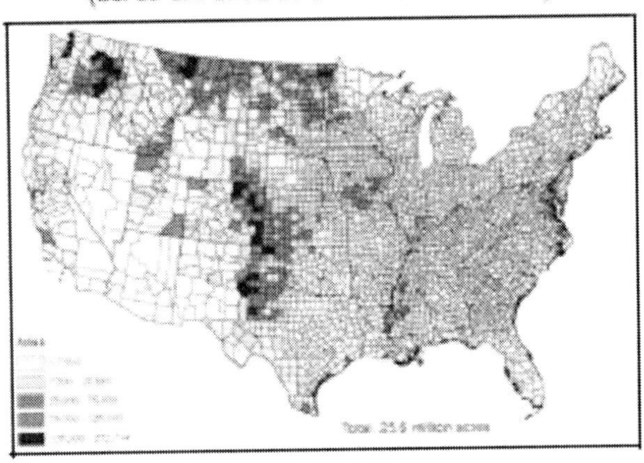

Source: USDA, FSA/EPAS/NRA, CRP Enrollment—November 2013, http://www.fsa.usda.gov/Internet/FSA_File/crpenrollmentnov2013.pdf.

Figure 3. CRP Enrollment.

Expiration and Reenrollment

CRP contracts vary in length, though most are 10 years in duration. At the end of a contract, the participant may either seek reenrollment into the program (via a general or continuous sign-up, if eligible) or let the contract expire. This 10-year cycle resulted in more than 16 million acres enrolled in 1997 potentially expiring all at once in 2007. To stagger this expiration process, USDA offered two- to five-year reenrollment and extension contracts in 2006 to contacts expiring between 2007 and 2010 (27 million acres). While approximately 83% of those offered accepted these extensions (23 million acres), over 8.5 million acres expired from the program during that time.

The 45th general sign-up in 2013 recorded 1.57 million acres of the 1.68 million acres deemed acceptable. Acceptable acres had an EBI score of 209 and above and an average rental rate of $64/acre. Approximately 3.3 million acres under CRP contract (both general and continuous) were scheduled to expire on September 30, 2013. Following the FY2013 general sign-up and continuous sign-ups, 1.7 million acres left CRP on October 1, 2013, either through contract expiration or attrition. An estimated 2 million acres are scheduled to expire at the end of FY2014. Between FY2007 and FY2014, over 17.1 million CRP acres under contact have expired and were not reenrolled in the program (*Figure 3).*

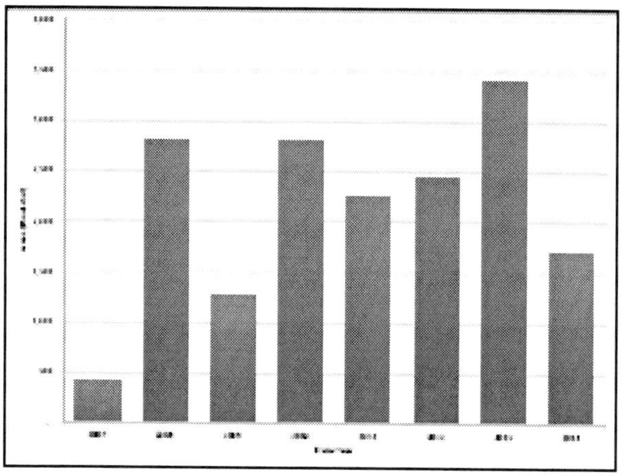

Source: USDA, FSA, Changes in CRP Acres from 2007 to October 2014 by State, http://www.fsa.usda.gov/Internet/FSA_File/acresstate2007102013.xls.

Notes: Annual sum of expiring land that was not reenrolled and land lost through attrition. FY2014 land may be cropped in the 2014 crop year.

Figure 4. CRP Acres Not Reenrolled and Lost Through Attrition, FY2007-FY2014.

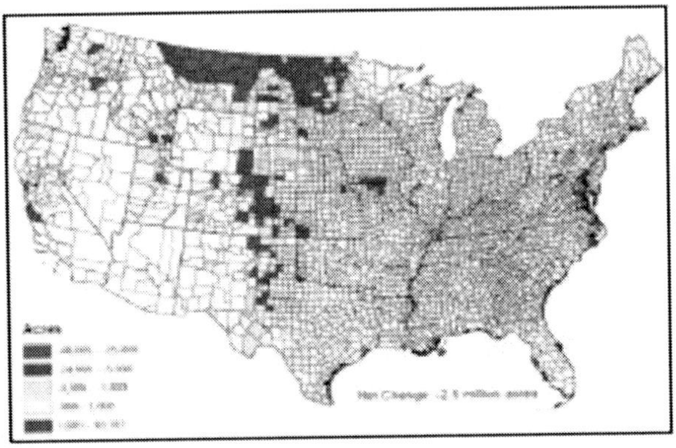

Source: USDA, FSA, EPA, NRA, CRP Change in Enrollment from September to October 2012, http://www.fsa.usda.gov/Internet/FSA_File/ acresmapsep toct 2012.pdf.

Notes: Contracts expire September 30; most new contracts begin October 1.

Figure 5. Change in CRP Enrollment Between September and October 2012.

The majority of these acres in FY2013 are located in the central part of the United States, which also has the largest number of acres enrolled (*Figure 5* and *Figure 3*). The number and location of these acres concerns some program advocates because of the potential loss of environmental benefits, particularly migratory and grassland bird habitat. Under the 1985 farm bill, CRP was initially authorized to enroll up to 45 million acres between crop years 1986 and 1990. USDA did not reach this enrollment cap and subsequent farm bills reduced the authorized level of enrollment.[14] CRP enrollment reached its peak in 2007 with 36.8 million acres. The 2008 farm bill reduced the enrollment cap to 32 million acres and the 2014 farm bill continues this reduction to 24 million acres by FY2018. This reduction in enrollable acres will further reduce the opportunity for reenrollment of expiring acres under contact and any new general sign-up would be relatively small.

Contract Termination and Early Release

Under current law, a producer wishing to terminate a CRP contract early faces a penalty of full repayment, with interest, of all the funds already paid to the producer, including any cost-share payments and other financial incentives, plus a fee of 25% of rental payments received. Although the Secretary of Agriculture always has the authority to release land from CRP without penalty, this option has not been commonly used. In program history, this option has been exercised twice—in 1995 and 1996—when acres were allowed a voluntary, penalty-free early release in order to enroll more environmentally sensitive cropland. In both cases, environmentally sensitive acres were not released and certain restrictions applied to acres returning to production or harvesting and grazing. In addition to the Secretary's discretion, an early-out provision has been in statute since the 1996 farm bill, but only applied to CRP contracts in effect before January 1, 1995. The 2014 farm bill amended this provision to allow for a one-time early release in FY2015, but requirements of the old provision remain, including:

- land cannot be devoted to filterstrips, waterways, strips adjacent to riparian areas, windbreaks, and shelterbelts,
- land cannot have an erodibility index of 15,
- rental payments must be prorated for the year of termination,
- land is still eligible for future CRP contracts, and
- land is still subject to conservation compliance requirements.

Section 2006 of the 2014 farm bill expands this provision to allow CRP contract holders to terminate prior to the contract's expiration in FY2015 if the contract has been in effect for at least five years and meets additional eligibility requirements. Specifically, in addition to the existing requirements above, the land is considered *ineligible* if it is

- devoted to hardwood trees,
- wildlife habitat, duck nesting habitat, pollinator habitat, upland bird habitat buffer, wildlife food plots, State acres for wildlife enhancement, shallow water areas for wildlife, and rare and declining habitat,
- a farmable wetland and restored wetland,
- land that contains diversions, erosion control structures, flood control structures, contour grass strips, living snow fences, salinity reducing vegetation, cross wind trap strips, and sediment retention structures
- located within a federally designated wellhead protection area,
- covered by an easement under CRP,
- located within an average width, according to the applicable NRCS field office technical guide, of a perennial stream or permanent water body, and
- enrolled under CREP.

Generally, most conservation and wildlife organizations are opposed to this penalty-free early-out option, however, they believe the additional restrictions on ineligible land could minimize the overall impact.[15]

Rental Rates

CRP rental payments are based on two main factors: the county average rental rate and soil productivity. The county average rental rate uses the National Agricultural Statistics Service's (NASS's) survey of county average rental rates for cropland and pastureland.[16] Soil productivity is based on a Natural Resources Conservation Service (NRCS) calculation that uses data of the local soil, landscape, and climate to determine the ability of the land to produce crops on non-irrigated soil. The average CRP rental payment rates are as follows: $51.09 per acre for general sign-up, $97.25 per acre for non-CREP continuous sign-up, $136.70 per acre for CREP continuous sign-up, and $111.70 per acre for farmable wetlands.[17]

Rental rates for CRP contracts became an important issue to some producers when commodity prices began to rise in 2008. Commodity prices have remained high, causing producers to claim that CRP rental rates are significantly lower than the producers could get by renting their land out for production. On the other hand, contracts are for 10 or more years and could be viewed as long-term investments rather than reactions to short-term commodity price fluctuation.

Economic Research Service Study

If rental rates are set too low, producers might decline to enroll their land, or, if already enrolled, they might decline to renew their contracts at expiration. A 2011 study by the Economic Research Service (ERS) at USDA modeled the effect of increasing commodity prices on CRP enrollment.[18] The study suggested that maintaining CRP under its current configuration could lead to program cost increases. When constraints were placed on increasing rental rates, the study suggested that enrollment goals could be met with moderate increases in the CRP rental rates. The latter scenario might mean that enrollment goals could be met, but at the cost of applying a lower EBI, as producers with profitable, but environmentally sensitive, acreage choose not to enroll.

If crop prices remain high and enrolling environmentally sensitive land continues to be a program priority, then finding the level of rental payments that encourages enrollment and keeps the cost of the program acceptable to policy makers might continue to be an issue.

Office of the Inspector General Report

In July 2012, the USDA Office of Inspector General (OIG) issued a report on the use of CRP rental rates.[19] The report concluded that for the 39th general sign-up in 2010, FSA (1) did not use the most recent NRCS soil productivity factors; and (2) allowed states and counties to propose alternate rates that did not adhere to its own policies for reviewing and approving the alternate rates.[20] OIG accepted two of the four agency responses to its recommendations. The two responses that it found unacceptable concerned the use of the most recent NRCS productivity data and establishing procedures for approving alternate county average rental rates. It is unclear what follow-up action, if any, will be taken by FSA to address the remaining concerns.

Increases in Enrollment

Despite the potential limiting factors affecting CRP enrollment, the number of acres enrolled under continuous sign-ups, including for the Conservation Reserve Enhancement Program (CREP, see *Appendix A*), has increased. Continuous sign-ups allow landowners to enroll land in certain high priority practices in exchange for additional financial incentive. As of July 2014, almost 5.8 million acres (23%) were enrolled through continuous sign-up, an increase of 2.2 million acres since 2007. The additional financial incentive under continuous sign-up could offset the potential gap between CRP rental payments and local rental rates to enroll more environmentally sensitive acres. More contracts and farms are enrolled under continuous sign-ups (409,713 and 239,033, respectively) than for general sign-ups (262,417 and 177,983, respectively).[21] Continuous sign-up enrollment represents a fraction of the total CRP acreage because, in general, these enrollments involve only a small portion of a farmer's total acreage.

Tax Status of Payments

Since the inception of CRP, the Internal Revenue Service (IRS) has issued a number of rulings, notices, and opinions on whether CRP rental payments are regarded as income from the "trade or business"[22] of farming and therefore subject to self-employment taxes of 15.3%. For years, the IRS generally treated CRP payments as farming income (subject to the self-employment tax) when received by someone who was engaged in the trade or business of farming, or as rental income (not subject to self-employment tax) when received by others (e.g., investors, heirs, and absentee landowners).[23]

The IRS recently changed its position, stating that by entering into a CRP contract the participant is in the trade or business of farming and therefore CRP payments are subject to the self-employment tax.[24] This was upheld in the courts in the 2013 U.S. Tax Court ruling of *Morehouse v. Commissioner*.[25] In the *Morehouse* decision, the Tax Court ruled in favor of the IRS, stating that the CRP participant (Morehouse) was in the business of maintaining "an environmentally friendly farming operation," making CRP payments subject to self-employment taxes. The decision is currently on appeal in the Eighth Circuit U.S. Court of Appeals.

The *Morehouse* decision concerns many producers and conservation advocates. Conservation groups fear that the tax will discourage CRP participation and possibly encourage contract cancellations. Others fear that the 15.3% tax will lower the value of CRP land. Some also argue that *Morehouse* could have broader implications for investor income, including cash rent landowners or possibly stock investors.[26]

Environmental Benefits

The greatest concern over a reduced level of CRP acres is a reduced level of environmental benefits. Since its inception, research has shown that CRP has contributed to reduced levels of soil erosion, water quality improvement, and wildlife habitat development. While these benefits vary across the country, some conclude that without CRP there could be additional environmental degradation from agricultural production.[27] *Table 3* includes a list of conservation practices applied on CRP land that is set to expire from the program between FY2015 and FY2017. It is unknown how many of the practices would expire as a result of acres not reenrolling in CRP due to the reduced number of authorized acres. It is also unknown whether these practices would be maintained without a CRP contract. Landowners may choose to continue these practices voluntarily or through other federal, state, or local assistance. In large part, the majority of practices that could be lost if allowed to expire would be grasslands, both native and introduced species, new and existing plantings.

According to FSA, since 2002, CRP has reduced soil erosion by 325 million tons from pre-CRP levels each year. Since the program's inception in 1986, CRP has reduced more than 8 billion tons of soil erosion. Through FY2010, CRP has enrolled more than 2 million acres in wetlands and over 2 million in buffers. Other annual conservation benefits include an equivalent of approximately 52 million metric ton net reduction in carbon dioxide (CO_2) from sequestration, reduced fuel use and nitrous oxide emissions avoided from no fertilizer use; more than 2 million acres of wildlife habitat established; and a reduction of about 607 million pounds of nitrogen and 122 million pounds of phosphorus.[28] From a wildlife perspective, it is estimated that CRP land produces over 13.5 million pheasants and 2.2 million ducks each year through habitat availability.[29]

Table 3. Top 10 Expiring CRP Conservation Practices (By expiring year in total acres applied)

Practice Code	Practice Description	2015	2016	2017
Total Expiring Acres		1,674,088	1,195,403	2,647,137
CP01	Introduced grasses and legumes	135,100	106,925	188,275
CP02	Native grasses	488,526	316,164	283,793
CP03H	Hardwood trees	59,385	5,388	31,800
CP04D	Permanent wildlife habitat	127,960	120,415	119,064
CP10	Existing grasses and legumes	164,116	81,697	1,353,896
CP11	Existing trees	39,479	36,817	112,723
CP21	Filter strips-grass	108,748	114,998	95,129
CP22	Riparian buffers	113,686	137,180	110,529
CP23	Wetland restoration	96,512	26,808	29,702
CP25	Rare and declining habitat	130,978	58,277	17,944

Source: CRS, developed from USDA, FSA data, April 2012.

CONCLUSION

The 113th Congress reauthorized CRP as part of the farm bill, but reduced the acreage cap to achieve a cost savings. Other pressures from high crop prices, increased demand for land, and the potentially low rental rates could also impact the program's ability to enroll the most desirable acres in the future. Despite these challenges, supporters encourage maintaining CRP enrollment because of the numerous environmental gains achieved by the program, including improved water quality, soil health, and wildlife species habitat. Balancing the cost of maintaining such benefits with the cost of the program could continue to be a challenge for Congress.

Appendix A. Continuous Sign-Up Initiatives

Continuous sign-up is designed to enroll the most environmentally sensitive land into CRP through specific conservation practices or resource needs. Continuous sign-up includes a number of initiatives that target acres with specific resource concerns or support additional conservation practices. These are described below.

Conservation Reserve Enhancement Program (CREP)

Initially implemented in 1997, CREP is a joint federal-state continuous sign-up program under CRP. CREP targets geographic regions with agricultural-related environmental concerns, such as Maryland's Chesapeake Bay and Florida's Everglades. Some states (e.g., New York and Ohio) have multiple CREP projects, each targeting a different area of the state. Projects are designed to address specific environmental objectives through targeted CRP enrollments. Sign-ups are continuous, non-competitive, and typically provide additional financial incentives beyond annual rental payments and cost-share assistance. There are currently 45 CREP agreements in 33 states.

Farmable Wetland Program (FWP)

The Farmable Wetland Program (FWP) enrolls farmable or prior converted wetlands into CRP. In exchange for additional financial incentives, landowners agree to restore the hydrology of the wetland, establish vegetative cover, and prohibit development. For land to be considered eligible it must meet one of the following criteria:

- a wetland or converted wetland cropped at least 3 of the immediately preceding 10 crop years;
- a constructed wetland that receives flow (surface and subsurface) from a row crop agriculture drainage system and is designed to provide nitrogen removal in addition to other wetland functions;

- land in a commercial pond-raised aquaculture in any year between 2002 through 2007; or
- cropland that was cropped at least 3 of 10 crop years between 1990 and 2002, and is subject to the natural overflow of a prairie wetland.

The enrollment of buffer acreage is also permitted to enhance wildlife benefits. No more than 100,000 acres may be enrolled in FWP in any state[30] and no more than 750,000 acres nationally. The enrollment of wetlands (described under the first and second bullets above) is limited to 40 contiguous acres. "Flooded farmland," or that defined in the fourth bullet above, is limited to 20 contiguous acres, and has a 20-acre limit.

Transition Incentive Program (TIP)

The 2014 farm bill reauthorized the Transition Incentive Program (TIP) option for expiring CRP contracts. Under TIP land from expiring CRP contracts may be transitioned back into sustainable grazing or crop production by a beginning, veteran, or socially disadvantaged farmer or rancher. The land must be from a retired or retiring owner or operator (not a family member) in exchange for up to two additional years of annual CRP rental payments following the expiration of the CRP contract. The program was authorized to spend up to $33 million between FY2014 and FY2018. This is an increase over the previously authorized $25 million in the 2008 farm bill.

Other Initiatives

Several other initiatives under CRP have been developed over time, mostly in response to Administration priorities. *Table A-1* includes a list of recent initiatives and their enrollment size.

Table A-1. CRP Initiatives

Initiative	Year Started	Allocation (acres)	Current Enrollment (acres)[a]
Floodplain Wetlands	2004	600,000	245,073
Bottomland Hardwood Trees	2004	250,000	93,922

Table A-1. (Continued)

Initiative	Year Started	Allocation (acres)	Current Enrollment (acres)[a]
Non-floodplain and Playa Wetlands	2005	350,000	264,824
Upland Bird Habitat Buffers	2005	500,000	243,738
Longleaf Pine Plantings	2007	250,000	119,927
Duck Nesting Habitat	2007	300,000	236,987
State Acres for Wildlife Enhancement (SAFE)	2008	1,250,000	845,837
Highly Erodible Land	2012	750,000	149,707
Pollinator Habitat	2012	100,000	5,499

Source: USDA, FSA, Conservation Reserve Program, Monthly Summary, July 2014, http://www.fsa.usda.gov/ Internet/FSA_File/julysummary14.pdf.

Notes: The 2014 farm bill limits the payment amount for thinning activities to no more than 150% of the total cost.

[a] Cumulative acres as of July 2014. Excludes lands enrolled in CREPs. Includes new and re-enrolled acreage under contracts that start in FY2015 (October 1, 2014).

APPENDIX B. CRP PROVISIONS IN THE 2014 FARM BILL

Table B-1 compares CRP provisions in the Agricultural Act of 2014 (P.L. 113-79) and prior law. U.S. Code citations are included in brackets in the 'Prior Law' column. Corresponding section numbers in P.L. 113-79 are included in brackets in the 'Enacted 2014 Farm Bill' column.

Table B-1. CRP Amendments in the 2014 Farm Bill Compared to Prior Law

Prior Law	Enacted 2014 Farm Bill (P.L. 113-79)
General Provisions	
Sec.1231(a-b) of the Food Security Act of 1985 (FSA) (P.L. 99-198, or the 1985 farm bill), as amended, authorizes the CRP through FY2013. CRP provides annual rental payments to producers to replace crops on highly erodible and	Extends authorization through FY2018. Adds grasslands to list of eligible lands, which is consistent with the consolidation of Grassland Reserve Program (GRP) rental agreements under CRP (also see Duties of the Secretary, Sec. 1233 of FSA). Amends

Prior Law	Enacted 2014 Farm Bill (P.L. 113-79)
environmentally sensitive land with long-term resource conserving plantings. [16 U.S.C. 3831(a-b)]	eligible land definition for land not enrolled in CRP to include filterstrips and land enrolled in other conservation practices. [Sec. 2001(a-b)]
Sec. 1231(c) of the FSA, as amended, determines the planting status of certain land. [16 U.S.C. 3831(c)]	Deletes language allowing land enrolled in the Water Bank Program and CRP cropland expiring in CY2000-CY2002 to be enrolled. [Sec. 2001(c)]
Sec. 1231(d) of the FSA, as amended, authorizes the maximum acreage enrollment levels; the program is authorized through FY2013 to enroll up to 32 million acres. [16 U.S.C. 3831(d)]	Reduces enrollment to 27.5 million acres in FY2014; 26 million acres in FY2015; 25 million acres in FY2016; 24 million acres in both FY2017 and FY2018. Within the total CRP acreage cap, grassland enrollment is limited to 2 million acres between FY2014-FY2018. Gives expiring CRP acres priority enrollment for grassland contracts. Grassland sign-up is continuous with one or more ranking periods. [Sec. 2001(d)]
Sec. 1231(e) of the FSA, as amended, defines the duration of contracts. [16 U.S.C. 3831(e)]	Amends language for land devoted to hardwood trees, shelterbelts, windbreaks, or wildlife corridors to allow flexible contract lengths beyond the current 10-15 years. [Sec. 2001(e)]
Sec. 1231(f) of the FSA, as amended, lists conservation priority areas as the Chesapeake Bay Region, the Great Lakes Region, and Long Island Sound. Watersheds with significant adverse water quality or habitat impacts related to agricultural production activities are eligible for priority designation. The areas' designation expires after 5 years or upon application of the state. [16 U.S.C. 3831f]	Deletes the watershed-specific language, but retains the use of conservation priority areas as determined by USDA. [Sec. 2001(f)]
Farmable Wetlands Program Sec. 1231B(a-f) of the FSA, as amended, authorizes a pilot program for up to 1 million acres of wetland and buffer acreage in CRP. [16 U.S.C. 3831b]	Renames the pilot program "Farmable Wetlands Program." Reauthorizes the program through FY2018, and clarifies language related to constructed wetlands receiving water from agricultural drainage. Reduces acreage limitation from 1 million acres to 750,000 acres. [Sec. 2002]
Duties of Owners and Operators	

Table B-1. (Continued)

Prior Law	Enacted 2014 Farm Bill (P.L. 113-79)
Sec. 1232(a)(8) of the FSA, as amended, establishes approved use of harvesting, grazing, and wind turbine use on CRP acres. Approved activities require a reduction in rental rate payments commensurate with the economic value of the authorized activity. [16 U.S.C. 3832(a)(8)]	Deletes language related to harvesting, grazing, and wind turbine use on CRP acres and adds similar language under the Duties of the Secretary section (Sec. 1233 of FSA). [Sec. 2003(a)]
Sec. 1232(b & d) of the FSA, as amended, requires a conservation plan on all CRP acres and reduces rental payment for certain authorized uses. [16 U.S.C. 3832(b & d)]	Amends conservation plan language by removing possible base acre retirement. Deletes rental payment reduction requirement for certain authorized activities and adds similar language under the Duties of the Secretary section (Sec. 1233 of FSA). [Sec. 2003(b-c)]
Duties of the Secretary	
Sec. 1233 of the FSA, as amended, specifies the duty of USDA to make cost-share payments and rental payments. [16 U.S.C. 3833]	Deletes the current section and adds new section that includes the approved use language from Duties of Owners and Operators section above (Sec. 1232 of FSA). In return for a CRP contract the Secretary must make cost-share and rental payments. Certain permitted activities are allowed if consistent with an approved conservation plan and are subject to restrictions for nesting birds that are economically significant, in decline, or conserved by law. Emergency harvesting, grazing, and other use of forage is permitted without a reduction in rental rate. Livestock grazing for a beginning farmer or rancher is permitted without a reduction in rental rate. Other certain permitted activities (harvesting, grazing, and wind turbines) are permitted in exchange for not less than a 25% reduction in rental rates. Grazing, harvesting, and fire suppression is permitted on enrolled grasslands. In exchange for a reduced rental rate, a landowner may install land improvement practices up to one year before the CRP acres expire. This land may not reenroll in CRP for five years. [Sec. 2004]
Payments	

Prior Law	Enacted 2014 Farm Bill (P.L. 113-79)
Sec. 1234 of the FSA, as amended, establishes a framework for calculating annual rental payments. [16 U.S.C. 3834]	Specifies that tree and shrub maintenance cost share payments are limited to between 2 and 4 years beginning on the date of planting. Adds the requirement that incentive payments be limited to no more than 150% of the cost of thinning or other practices conducted. Amends rental payment calculation to include grassland contracts for not more than 75% of the grazing value. Adds the requirement that NASS conduct a rental rate survey no less than once a year. Dryland cash rental rates may also be used as a factor for determining annual rental rates. Deletes language allowing for in-kind commodities as a form of CRP payment. Payments must be made in cash and may be in advance of performance determinations. [Sec. 2005] Sec. 2601(a) includes a limit of $10 million for thinning activities between FY2014-FY2018.
Contract Requirements	
Sec. 1235(e) of the FSA, as amended, allows owners and operators to terminate a contract entered into before January 1, 1995, at any time if the contract has been effect for at least 5 years. Land with filterstrips, waterways, strips adjacent to riparian areas, windbreaks, shelterbelts, erodibility index of more than 15, and other land of high environmental value (e.g., wetlands) are not eligible for early release. The contract termination becomes effective 60 days after the participants notice. Rental payments are prorated and conservation compliance requirements remain in effect. [16 U.S.C. 3835(e)]	Allow owners and operators to terminate their CRP contracts in FY2015 if the contract has been in place for at least 5 years. Adds to the list of excepted land, including land with: hardwood trees, wildlife habitat, duck nesting habitat, pollinator habitat, upland bird habitat buffer, wildlife food plots, State Acres for Wildlife Enhancement (SAFE), shallow water areas for wildlife, rare and declining habitat, farmable wetlands, restored wetlands, diversions, erosion control structures, flood control structures, contour grass strips, living snow fences, salinity reducing vegetation, cross wind trap strips, sediment retention structures, federally designated wellhead protection areas, an easement under CRP, and average width of a perennial stream or permanent water body, and a CREP contract. Terminations become effective upon approval. [Sec. 2006(a)]
Sec. 1235(f) of the FSA, as amended,	Adds "veteran farmer or rancher" as eligible

Table B-1. (Continued)

Prior Law	Enacted 2014 Farm Bill (P.L. 113-79)
facilitates the transfer of CRP acres from a retiring owner to a beginning/socially-disadvantaged producer to return land to production, and allows new owner to begin land improvements or start organic certification process one year before CRP contract expires. [16 U.S.C. 3835(f)]	individuals for the transition option, in addition to beginning farmer or rancher. Specifies that approved land improvements include preparing to plant an agricultural crop. [Sec. 2006(b)] Reauthorizes and increases the limit on the CRP transition option to $33 million for FY2014-FY2018. [Sec. 2601(a)]
No comparable provision	Adds a provision allowing landowners to enroll in the Conservation Stewardship Program (CSP) and conduct activities required under CSP in the final year of the CRP contract without violating the terms of the contract. Also allows USDA to terminate or modify a CRP contract if eligible land is transferred into the Agricultural Conservation Easement Program (ACEP). [Sec. 2006(c)]
Sec. 1235A of the FSA, as amended, allows land enrolled in CRP before enactment of the 1990 farm bill (P.L. 101-624, November 28, 1990) to convert vegetative cover to hardwood trees or restored wetlands [16 U.S.C. 3835a]	Repeals provision. [Sec. 2007]
No comparable provision.	Provides transition language stating that changes made by the 2014 farm bill do not affect the validity or terms of existing contracts. Allows CRP participants to update their current contract to reflect the new terms and conditions under the Sec. 2004—permitted activities. [Sec. 2008]
Funding	
Sec. 1241(a)(1) of the FSA, as amended, allows the use of funds, facilities, and authorities of the Commodity Credit Corporation to carry out CRP. Limits payments for thinning activities to $100 million between FY2009-FY2013 and payments for the transition assistance (see Sec. 1235(f) above) to $25 million for FY2009-2013. [16 U.S.C. 3841(a)(1)]	Reduces limit for incentive activities (see Sec. 2005) to $10 million between FY2014-FY2018 and increases limit for transition assistance (see Sec. 2006) to $33 million between FY2014-FY2018. [Sec. 2601(a)]

Source: CRS.

End Notes

[1] Amendments in the 2014 farm bill reduces the enrollment cap to 26 million in FY2015, 25 million in FY2016, and 24 million in FY2017 and FY2018.

[2] USDA, FSA, Conservation Reserve Program Sign-Up 45 Environmental Benefits Index (EBI), Fact Sheet, February 2013, http://www.fsa.usda.gov/Internet/FSA_File/ su45ebifact sheet. pdf.

[3] USDA, FSA, Conservation Reserve Program, Monthly Summary, July 2014.

[4] U.S. Department of Agriculture, "USDA Announces Programs to Conserve Sensitive Land and Help Beginning Farmers," press release, June 4, 2014, http://www.usda.gov/wps/portal/usda /usdahome?contentidonly=true&contentid= 2014/06/0112.xml.

[5] Both CREP and the farmable wetlands programs were reauthorized in the 2014 farm bill and are discussed in the Appendix A.

[6] 7 C.F.R. 1410.5. Producers must also meet broader eligibility requirements related to adjusted gross income limits (not more than $900,000) and compliance requirements.

[7] 16 U.S.C. 3831(b).

[8] Refers to a conservation plan developed under the highly erodible land conservation provisions. For additional information, see CRS Report R42459, Conservation Compliance and U.S. Farm Policy, by Megan Stubbs.

[9] 27.5 million in FY2014, 26 million in FY2015, 25 million in FY2016, and 24 million in FY2017 and FY2018., 16 U.S.C. 3831(d).

[10] For example, the highest average rental payment per acre for a state (for all CRP sign-ups) is in Massachusetts at $207.20/acre. Only 10 acres are enrolled in Massachusetts. The lowest is in Wyoming at $26.57/acre. The state with the most acres enrolled—Texas at 3.2 million acres—averages $36.78/acre for rental payments.

[11] Includes vegetation management requirements and restricted periods of application. Both requirements are developed in consultation with State Technical Committees.

[12] The easement portion of GRP was incorporated into the new Agricultural Conservation Easement Program established in the 2014 farm bill. The repeal does not affect the validity or terms of any GRP contract or easement entered into prior to the date of enactment (February 7, 2014) or any associated payments required to be made in connection with an existing contract or easement.

[13] Activities were subject to certain restrictions during nesting seasons of bird species that were in significant decline or were protected under Federal or State law. All participants were required to operate under an approved grazing management plan.

[14] The 1996 farm bill (P.L. 104-127) lowered the enrollment cap to a total of 36.4 million acres through CY2002. The 2002 farm bill (P.L. 107-171) increased the enrollment cap total to 39.2 million acres through CY2007.

[15] National Wildlife Federation, 2014 Farm Bill Conference Report Analysis, 2014, p. 3, http://www.nwf.org/~/media/ PDFs/wildlife/farm%20%20bill/NWF-2014-Farm-Bill-Analy sis_final.pdf.

[16] The 2014 farm bill amended the requirement that NASS conduct a survey no less than once a year on county average market dryland and irrigated cash rental rates. USDA may uses the survey estimates relating to dryland cash rental rates when determining annual rental rates.

[17] USDA, FSA, Conservation Reserve Program - Monthly Summary, Washington, DC, July 2014.

[18] Daniel Hellerstein and Scott Malcolm, The Influence of Rising Commodity Prices on the Conservation Reserve Program, USDA, ERS, Economic Research Report number 110, February 2011.

[19] USDA, OIG, Farm Service Agency, Conservation Reserve Program - Soil Rental Rates, Audit Report 03601-0051- Te, July 2012.

[20] For example, "of the 687 proposed alternate rates it received, FSA's national office approved 686, even though it determined the majority (669 of 687) of the proposals contained evidence to support the alternate rates to be less than strong." Under the 41st general sign-up, a total of 271 proposals were submitted for alternate county average rental rates. According to OIG, one official approved 150 rates—105 exceeded the NASS rate and 45 were lower than the NASS rate. Ibid.

[21] USDA, FSA, Conservation Reserve Program, Monthly Summary, October 2013, http://www.fsa.usda.gov/Internet/ FSA_File/oct2013summary.pdf.

[22] 26 U.S.C. 1402(a) excludes "rentals from real estate" from the definition of "net earnings from self-employment," which includes the gross income derived by an individual from any "trade of business" carried out by the individual, less deductions.

[23] An exemption was created in the 2008 farm bill (P.L. 110-246) which provides a statutory exemption from self-employment taxes provided for CRP payment recipients who are also collecting Social Security or disability payments (Section 15301 of P.L. 110-246).

[24] C.C.A. 2003-25-002 (May 29, 2003) and I.R.S. Notice 2006-108, 2006-51 I.R.B. 1118.

[25] Morehouse v. Commissioner, 140 T.C. No. 16 (2013).

[26] Roger A. McEowen, U.S. Tax Court Says Non-Farmer's CRP Income is Subject to Self-Employment Tax, Center for Agricultural Law and Taxation, Ames, IA, June 20, 2013, http://www.calt.iastate.edu.

[27] JunJie Wu and Bruce Weber, Implications of a Reduced Conservation Reserve Program, The Council on Food, Agriculture & Resource Economics, July 2012.

[28] USDA, FSA, The Environmental Benefits of the Conservation Reserve Program, United States—2010, July 2011, http://www.fsa.usda.gov/Internet/FSA_File/united_states.pdf.

[29] National Wildlife Federation, Maintaining Benefits of Expiring CRP, http://www.nwf.org/Wildlife/Policy/Farm-Bill/ Farm-Bill-Success-Stories/Success-Expiring-CRP.aspx.

[30] This limit may be increased by USDA to up to 200,000 acres per state.

In: The Conservation Reserve Program
Editor: Preston L. Cruz
ISBN: 978-1-63482-282-4

Chapter 2

THE INFLUENCE OF RISING COMMODITY PRICES ON THE CONSERVATION RESERVE PROGRAM*

Daniel Hellerstein and Scott Malcolm

ABSTRACT

This report considers how increased commodity prices might influence enrollment in and benefits from the Conservation Reserve Program (CRP) using two complementary models: a *likely-to-bid* model that uses National Resources Inventory data to simulate offers to the general signup portion of the CRP and an *opt-out* model that simulates retention of current CRP contracts. Under several higher crop price scenarios, including one that incorporates 15 billion gallons of crop-based biofuels production, maintaining the CRP as currently configured will lead to significant expenditure increases. If constraints are placed on increasing rental rates, it might be possible to meet enrollment goals with moderate increases in CRP rental rates—but this will mean accepting lower average Environmental Benefits Index scores as landowners with profitable but environmentally sensitive lands choose not to enroll.

---

* This is an edited, reformatted and augmented version of Economic Research Report Number 110, issued by the U.S. Department of Agriculture, Economic Research Service, February 2011.

Keywords: CRP, ethanol, commodity prices, likely to bid, REAP

SUMMARY

What Is the Issue?

The Conservation Reserve Program (CRP) is a voluntary program that enables farmers to retire farmland for conservation in exchange for rental payments. After a period of relative stability, the CRP is now facing a number of changes. The 2008 Farm Act reduced the CRP's maximum enrollment to 32 million acres—4.6 million acres less than the program's peak acreage in 2007. Moreover, higher commodity prices since 2006 are likely to lead to increases in cropland rental rates and inflate CRP program costs. If high prices become the norm, landowner interest in CRP may wane as they weigh the expected returns to farming against the CRP payment, particularly if CRP rental rates do not keep up with market rental rates. This could lead to fewer acres being offered to the program, with a commensurate drop in ecosystem services.

What Did the Study Find?

Using a computer simulation, the authors of this report demonstrate that, in an era of elevated crop prices, maintaining the CRP's acreage, and the environmental benefits it provides, will require higher program payments.

- If commodity prices and CRP rental rates prevalent in 2007 were maintained over the long term (commodity prices that are higher than when most CRP contracts enrolled) the quantity and quality of land offered to the program would decline. An increase in these 2007 rental rates by 60 percent would largely offset the long-term impact of the higher prices, albeit with a corresponding increase in program costs.
- Given the established interest in the program and its longstanding popularity with landowners, if USDA's policy of using prevailing rental rates were altered, it might be possible to meet acreage goals with moderate increases in payment rates. But this would mean accepting offers providing fewer environmental benefits, as

landowners with environmentally sensitive, but increasingly profitable, lands choose to withhold their land from the program.

- Any additional impacts caused by an increase in crop-based ethanol production from 6.5 to 15 billion gallons per year would be relatively minor. For example, given a 60-percent increase in CRP rental rates, the model predicts 2 million fewer acres (about 5 percent of the current total) offered to the program over the long term and a 2-percent reduction in environmental benefits (as measured by the CRP's Environmental Benefit Index). When considering currently enrolled CRP acres, this expansion in ethanol production leads to about 200,000 acres leaving the program.
- Using the unusually high crop prices seen in summer 2008, the model shows a large response by CRP participants. Maintaining the CRP payments at their current level results in fewer acres offered to the program, making it unlikely that the program could reach its goal of 32 million acres. Over the long term, to enroll acreage that would maintain the environmental benefits currently provided by the program would require roughly doubling CRP rental rates.
- If a robust carbon market permitted all CRP enrollees to also sell carbon offsets from their retired land, the impacts of increased commodity prices on the costs of CRP could be substantially reduced.

How Was the Study Conducted?

We use two modeling strategies. A *likely-to-bid* model predicts what the CRP would look like if the program were to start from scratch and enroll the entire acreage in one hypothetical signup— it is a steady-state model that starts with postulated prices, which are assumed to stay constant, and predicts what the CRP will look like over the long term. An *opt-out* model treats current acreage as a given, and predicts which acres would be withdrawn and converted to which crops under different price and rental payment levels.

Each model has inherent strengths and limitations that are partially compensated for in the other approach.

- The Economic Research Service's *likely-to-bid* model (LTB) is a simulation model that predicts how representative parcels of land will respond to a CRP program under varying circumstances. As part of the methodology, the LTB predicts factor scores for the Environmental Benefits Index (EBI).

- Contract data on all currently enrolled CRP acreage is the basis of the *opt-out* model. Contract data contain for each contract the actual value of CRP payments, the maximum payment permitted, and the EBI factors.
- Several scenarios that illustrate possible market and program situations are constructed. The scenarios incorporate moderate and high price increases over historical trends. For scenarios that consider the impacts of increasing ethanol production to 15 billion gallons per year, the Regional Agriculture and Environment Programming model (REAP) was used to determine prices and crop shares. The mitigating effects of increasing CRP rental payments are included as another factor in the scenarios. Offered and accepted acres, average Environmental Benefits Index scores, forgone agricultural revenue, average rental payments, and regional distribution of CRP acreage are computed for each scenario.

INTRODUCTION

The Conservation Reserve Program (CRP) is one of the USDA's largest environmental programs. Now over 25 years old, the CRP is a voluntary program that pays landowners to convert environmentally sensitive cropland to grass, tree, and wetland covers. CRP contracts cover a multi-year period, with significant penalties for withdrawing before the contract expires. Originally established to protect soil quality, the CRP's primary goals now include water and air quality protection and the enhancement of wildlife habitat.[1]

The CRP uses competitive bidding to select which lands to enroll in its general signup. Since landowner interest in the CRP is influenced by a number of external factors, the makeup of the CRP will change as market forces, and program policies, change. For example, a jump in commodity prices could reduce acreage offers, both new and renewing. On the other hand, if the USDA increases per-acre rental rates, it could counteract the effects of increased commodity prices. Or the USDA could change its ranking criteria and alter the types of land accepted into the program. In either case, the regional distribution of CRP acreage, the types of benefits these acres produce, and the cost of the program would likely change.

The goal of this report is to investigate what could happen to the CRP under alternative assumptions about commodity price increases and alternative assumptions about management of the programs. We focus on several questions, including:

- How might high prices affect the makeup of the CRP? The past few years witnessed a boom in agricultural commodity prices—a boom that corresponded with the growth in the production of biofuels. Even though prices have dropped from their summer 2008 peaks, they are still substantially higher than when most existing CRP contracts were signed
- How might further increases in the production of corn-based ethanol affect the CRP?
- How would higher CRP rental rates affect program expenditures and the types of land enrolled?
- How will the reduction in the size of the CRP affect the program? The 2008 Farm Act capped the CRP at 32 million acres, versus nearly 37 million acres in 2007. In reducing the size of future signups or allowing enrolled acres to opt out, which regions or classes of land are likely to be less prominent in CRP?
- How might offset markets for carbon dioxide (CO_2) affect CRP enrollment and costs?

THE CONSERVATION RESERVE PROGRAM: BACKGROUND

The Conservation Reserve Program (CRP), established by the Food Security Act of 1985, enables eligible landowners to remove land from crop production in exchange for a fixed payment over a fixed period. By 1990, 32.8 million acres were enrolled in 10- or 15-year contracts. Initially, the criteria for enrollment were largely based on soil erosion potential, and nearly all acres that fulfilled crop history and other eligibility requirements were accepted.[2]

By 1997, many of the early CRP contracts were expiring, leading to several large general signups that enrolled 27 million acres through 1999. Unlike the initial signups, CRP offers were ranked on several criteria, with soil erosion, water quality, and wildlife habitat of equal importance. A newly created Environmental Benefits Index (EBI) used a number of factors—including soil erodibility, the land cover planted, and proximity to water bodies—to rank each offer. A cost factor was included that rewarded offers less costly to USDA.[3] During these latter signups, interest in the program (as measured by total acres offered) exceeded enrollment limits. In 1997, only 70 percent of offers were accepted.[4]

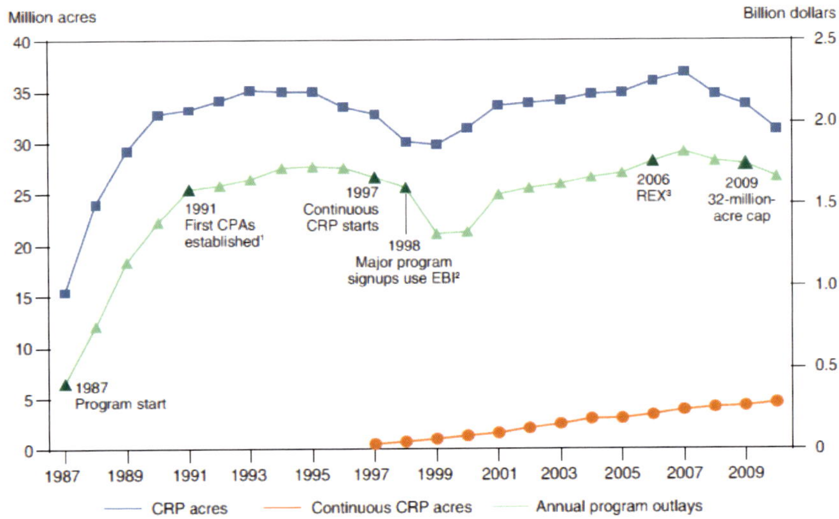

Source: USDA, Economic Research Service, using CRP contract data maintained by USDA's Farm Service Agency.

[1] CPA = Conservation Priority Areas

[2] EBI = Environmental Benefits Index

[3] REX = Re-enrollment and extension program.

Figure 1. History of CRP: Acreage and expenditures, with important events.

The decade following the large general signups of 1997 witnessed CRP acreage fluctuating between 30 and 37 million acres, with signups ranging from 1 million to 4 million acres per signup. The EBI continued to be used, with minor modifications. A more significant change was the advent of "continuous CRP." Starting in 1996 and now comprising about 4 million acres, continuous CRP accepts lands that meet more stringent eligibility criteria, both in terms of location and cover crop. Acres can be enrolled in a continuous signup at any time and are noncompetitive (if an offer is eligible, it is automatically accepted). In addition, the program tends to both pay higher rental rates and offer targeted incentives for continuous CRP acres.

Program expenditures have been fairly steady throughout the program's history, increasing from $1.5 billion in 1990 to $1.9 billion in 2007 (in nominal dollars). The CRP's per-acre rental payments are closely related to non-irrigated cropland rental rates, which were relatively stable in the years following 1997.[5]

However, beginning in 2006 there were several changes that influenced the program. First, over 16 million acres of CRP land enrolled in 1997 were set to expire in 2007. For several reasons, including administrative efficiency, the USDA sought to stagger the timing of these expirations. Hence, in 2006 USDA offered 2- to 5-year re-enrollments and extensions to owners of contracts covering 27 million acres scheduled to leave the program between 2007 and 2010. Contract holders covering about 23 million acres (83 percent) accepted.

Second, the Food, Conservation, and Energy Act of 2008 (2008 Farm Act) capped CRP enrollment at 32 million acres. This was achieved as of October 1, 2009, by permitting approximately 2 million acres in expiring contracts to leave the program without an offer to renew or extend.

Last, starting in 2007 agricultural prices rose significantly, reaching a peak in the summer of 2008.[6] For example, corn prices jumped from $3.03 in the 2006 market year to $4.20 in the 2007 market year, and then further increased to $5.40 in the summer of 2008. Although prices have subsequently moderated, the factors underlying the increase have not disappeared.[7]

Increasing Commodity Prices and the CRP

Given USDA predictions[8] of high commodity prices for the next several years (relative to price levels throughout the CRP's history), the opportunity costs of retiring cropland are likely to remain high. This could dampen land-owner willingness to enroll in the CRP's general signup (SWCS, 2008).

Several authors have considered the price-enrollment relationship. Secchi et al. (2009) used the EPIC model to estimate the environmental impacts of changes in CRP enrollment in Iowa due to increasing commodity prices. They conclude that under higher commodity prices, "maintaining current levels of environmental quality will require substantially higher spending levels," and argue for a targeted approach. To counteract the risk of higher crop prices' eroding the CRP's ability to protect ecologically sensitive lands, Baker and Galik (2009) argue for several program modifications to increase income flows from CRP lands, such as allowing the sale of carbon offset credits.[9] Although the commodity prices have dropped from the peaks reached in the summer of 2008, the impact of higher prices on the CRP's effectiveness and attractiveness is still a concern. This report investigates the implications of probable USDA actions (such as increasing rental rates), and potential USDA modifications to the program, such as selective retention of current CRP lands, and the consequences of an active market for carbon payments.

MODELING CRP ENROLLMENT

To examine how enrollment for the CRP general signup might change as prices and other factors change, we use two modeling strategies. Our *likelyto-bid* (LTB) model predicts what the CRP's general signup acres would look like if the program were to start from scratch and enroll the entire acreage in one hypothetical signup. Our *opt-out* model treats current general signup acreage as a given and predicts which acres would withdraw given the chance of opting out.

- The *likely-to-bid model (LTB)* (see appendix 1, "Likely-To-Bid Model: Methodology") is a simulation model that predicts how representative parcels of land will respond to a CRP program under different circumstances. The LTB is a steady-state model: it starts with a postulated set of parameters (such as commodity prices, crop yields, and production costs), holds them constant, and predicts the long-term distribution of CRP acreage. It also predicts the environmental impacts of these acres, using predictions of factor scores in the CRP's Environmental Benefits Index (EBI).
- The *opt-out model* (see appendix 2) uses contract data on all CRP acreage, which contain for each contract the actual value of CRP payments, the maximum payment permitted, and the EBI factors.

In this analysis, we focus on acreage that will enter the program through general signups, and abstract from acreage enrolled via continuous signups. Since continuous signups offer higher rental payments than general signups, continuous signup acreage is less likely to be influenced by changes in commodity prices.

Both models have their strengths and weaknesses. The LTB model is national in scope, with disaggregated measures of land quality and other agriculturally important factors. However, its weaknesses include a dependence on increasingly dated data (much of it from 1997), the use of approximation when assigning EBI factor scores to representative parcels, and the use of imprecise rules to determine which representative parcels would actually be offered to the program.

The *opt-out* model uses actual CRP general signup contracts, which include precise measures of the EBI factors and rental rates. Modeling is based on actual rather than simulated participants. However, there are very limited data on land quality, which complicates estimates of the profitability of alternative uses of the land. And this dataset has no information on non-CRP land parcels.

Scenarios for Analysis

Our analysis of possible changes to CRP general signup enrollment is based on four scenarios (table 1).

- A *Baseline* that captures conditions when most CRP contracts were enrolled. This is a "one-shot" simulation that uses 2005 crop prices in order to calibrate the LTB model (it is not associated with USDA's published baseline).
- A *MediumPrice* scenario that reflects moderate price levels—we use prices prevalent in 2007.[10]
- A stylized *RFS (Renewable Fuels Standard)* scenario that captures price impacts of increasing ethanol production from 6.5 billion gallons to 15 billion gallons.[11]
- A *HighPrice* scenario that reflects a high price regime—using prices prevalent in the summer of 2008.

These scenarios were developed based on historical commodity prices and may or may not reflect USDA's current price projections.[12] They are meant to illustrate the relationships between commodity prices and CRP enrollment and should not be considered projections of future trends. The commodity prices used in the scenarios increase as one moves from the *Baseline* to the *HighPrice* scenarios. For example, corn prices are almost 190 percent higher in the *MediumPrice* scenario than in the *Baseline*, and 140 percent higher in the *HighPrice* than in the *MediumPrice* scenario (table 2). Some commodity prices in the RFC scenario are somewhat higher than those in the *MediumPrice* scenario—for example, corn price is 12 percent higher, but soybean prices are basically unchanged.

For each scenario, two variants are analyzed: one that uses 2007 CRP rental rates and a second that captures CRP rental rates increases (given changes in cropland rental rates) via an across-the-board increase in these 2007 rates.

Given the 2008 Farm Act's changes to the CRP, all scenarios use a target of 30 million general signup acres. As of early 2009, there were about 33.6 million acres of CRP land: 29.5 million in general signup and 4 million in continuous signup. By fall 2009, the CRP's enrollment had shrunk to around 31 million acres (with 27 million in general signup). Thus, although the use of a 30-million-acre target may overstate the size of CRP's general signup over the next several years, it is predicated on past levels of enrollment and allows for consideration of alternative future targets.

To compare scenarios, the following set of metrics is used.

- Offered acres. In general, the more acres that are offered, the better the choices available to CRP administrators.
- Average environmental score per accepted acre (EEBI). The EEBI is the EBI without the cost factor. While not an ideal measure of the real environmental impact of CRP land retirement, it captures several environ mental attributes and applies weights whose magnitudes reflect the value CRP administrators place on different environmental services/attributes.
- *Forgone net agricultural revenue.* This is a measure of the social costs of the program in terms of lost agricultural production.
- *Rental payments (total and per acre).* This is the cost of the program to the Government.
- *Regional distribution of CRP acres.* This is how CRP acres occur across the 10 farm production regions (http://www.ers.usda.gov/ briefing/arms/ resourceregions/resourceregions.htm).

Table 1. Description of scenarios

The following describes the scenarios used in the LTB and opt-out models. Table 2 details the commodity prices used in these scenarios.

Scenario	Description
Baseline	This scenario serves as a baseline that reflects circumstances that prevailed before the recent

Scenario	Description
	increase in crop prices. Specifically, this scenario uses commodity prices, production costs, soil rental rates, and EBI factors extant in 2005. The baseline captures historical conditions, such as measures of environmental benefits, that one can contrast with what the CRP would provide as prices and policies change. Furthermore, comparison of baseline results to actual CRP offers and contracts indicates the accuracy of the LTB model.
MediumPrice	This scenario reflects a continuation of current conditions. Specifically, its uses commodity prices, production costs, CRP rental rates, and EBI factors extant in 2007. Note that 2007 prices are similar to prices observed in spring 2010 (as are rental rates and EBI factor scores). Furthermore, prices in 2007 incorporate the impacts of 6.5 billion gallons of ethanol production.
MediumPrice variant: higher rental rates *(MediumPrice adjusted)*	Same as the *MediumPrice* scenario, but includes an across-the-board increase in CRP rental rates of 60%. This represents a possible response by USDA to increased commodity payments, hence increased non-irrigated cropland crop rental rates.
Increase crop-based ethanol production to 15 billion gallons *(RFS)*	To consider impacts of increased ethanol production as mandated by the Renewable Fuel Standard (RFS), we use the REAP (BRDi, 2008) model to generate estimates of prices that may prevail if crop-based ethanol production increases from 6.5 billion to 15 billion gallons. These are combined with production costs, EBI factor scores, and CRP rental rates used in the *MediumPrice* scenario.
RFS variant: increased rental rates *(RFS adjusted)*	Same as *RFS*, but includes an across-the-board increase in CRP rental rates of 60%.
HighPrice	This scenario reflects a possible high-price environment. It is based on prices observed in summer 2008, along with production costs, EBI factor scores, and CRP rental rates used in the *MediumPrice* scenario.

Table 1. (Continued)

Scenario	Description
HighPrice variant: much higher rental rates (HighPrice large adjusted)	Same as *HighPrice*, but includes an across-theboard increase in CRP rental rates of 120%.

Notes: LTB = likely to bid, EBI = Environmental Benefits Index, REAP = Regional Environment and Agriculture Programming model, RFS = Renewable Fuel Standard.

Table 2. Prices used in scenarios (and their variants)

Scenario	Corn	Sorghum	Wheat	Soybeans	Cotton
			$/bu		$/bale
Baseline	2.00	1.86	3.42	5.66	208.00
Medium Price	3.39	3.21	6.08	9.00	254.00
RFS	3.80	3.32	6.13	9.03	257.00
High Price	5.40	4.90	7.25	12.25	364.00

Notes: The same prices are used in the adjusted and large adjusted variants of each scenario. RFS = Renewable Fuels Standard.

INFLUENCE OF PRICES: LIKELY-TO-BID MODEL

We begin by comparing the performance of the LTB model with the actual data from CRP contracts and offers (table 3). For the *Baseline* scenario, the LTB predicts that 51.9 million acres would be offered (with a hypothetical CRP signup from scratch), which is quite close to the 49.2 million acres offered between 1997 and 2007.[13] Thus, the model's ability to replicate observed results is qualitatively reasonable, though not precise.

- The *Baseline* overpredicts EEBI scores by about 10 percent, but does a good job of predicting actual contract rental rates.
- Comparison of the regional distribution of acres (table 4) indicates that the rank order of regions identified by the *Baseline* is close to the actual order.
- The *Baseline* indicates that, on average, over 35 percent ($19/acre) of an offer's estimated rent ($47/acre average) is a premium over the LTB model's predicted net returns ($28/acre on average) from that

parcel. Kirwan et al. (2008) find that estimated premiums constitute 10-40 percent of the program's rental payouts. Since the LTB assumes no competitive bidding (all landowners request their maximum rental payment), it is not surprising that the model estimates a premium toward the upper end of this range.

When judging the validity of the *Baseline* model, one must consider that both contracts and offers are spread over a 10-year period (1997-2007). Over this time span, commodity prices, input costs, and EBI criteria were mostly stable, but did change in minor ways. Thus, it would be unlikely for the *Baseline* (a "one-shot" simulation, using 2005 prices) to precisely match any summary statistic on actual contracts or offers.[14]

Table 3. CRP general signup contracts and offers: actual and likely-to-bid predictions under the Baseline scenario

		Acres	EEBI score and normalized[4]	Forgone net agricultural revenue	Rental payments
		Million	Average/acre	Average $/acre	
Actual	Contracts[1]	32.6	173 (0.43)	n.a.	45
	Offers[2]	49.2	150 (0.37)	n.a.	45
Baseline	Contracts[3]	30.0	188 (0.48)	28	47
LTB	Offers[3]	51.9	165 (0.42)	32	51

Source: USDA/Farm Service Agency contract and offer files, and likely-to-bid (LTB) simulation results.

n.a.-Not applicable.

[1] General signup contracts as of March 2008.

[2] The Offers row aggregates actual offers to the CRP that were received in general signups 15 through 33 (conducted between 1997 and 2007).

[3] Both LTB rows are predicted values derived from the LTB model under the *Baseline* scenario.

[4] Since the Environmental Benefits Index (EBI) factor weights varied over time (for example, the N6 "CPA" factor was dropped after signup 20), the normalized EEBI score is used. This is the EEBI score divided by the maximum number of EEBI points available in that signup. For the LTB simulations, the signup 26-33 maximum (395) is used as the normalizing value.

Table 4. Geographic distribution of general signup CRP acres: actual and likely-to-bid (LTB) predictions under the Baseline scenario

		NE	AP	SE	DE	CB	LS	NP	SP	MT	PA
		Share of acres									
	Contracts	0.3	1.8	2.8	3.4	12.4	7.3	25.2	15.7	23.8	7.1
Actual											
	Offers	0.4	2.2	3.5	3.3	12.6	7.7	27.7	15.9	21.0	5.9
Baseline	Contracts	1.0	4.4	4.9	3.2	13.5	12.5	26.9	10.5	18.1	1.5
LTB	Offers	1.3	4.4	3.5	2.7	12.7	13.4	32.4	10.6	14.6	4.2

Source: USDA/Farm Service Agency contract and offer files, and LTB simulation results.

Note: Due to rounding, rows may not add to 100 percent.

NE = Northeast: New Hampshire, Pennsylvania, Maine, Maryland, Rhode Island, Massachusetts, Delaware, Connecticut, Vermont, New York.

New Jersey.

AP = Appalachia: West Virginia, Tennessee, North Carolina, Virginia, and Kentucky.

SE = Southeast: South Carolina, Alabama, Georgia, and Florida.

DE = Delta: Louisiana, Arkansas, and Mississippi.

CB = Corn Belt: Ohio, Iowa, Missouri, Indiana, and Illinois.

LS = Lake States: Minnesota, Michigan, and Wisconsin.

NP = Northern Plains: North Dakota, South Dakota, Kansas, and Nebraska.

SP = Southern Plains: Texas, Oklahoma.

MT = Mountain: Colorado, Utah, Arizona, New Mexico, Wyoming, Nevada, Idaho, and Montana.

PA = Pacific: Oregon, California, and Washington.

Table 5. LTB predictions of acres offered, and accepted, in the CRP general signup MediumPrice, RFS, and HighPrice scenarios

Scenario	Offered Acres	EEBI score[1] and normalized	Forgone net agricultural revenue[1]	Rental payments[1]
	Million	Average/acre	----Average $/acre----	
MediumPrice	28.8	161	34	64
MediumPrice adjusted	44.7	179	48	83
RFS adjusted	42.7	176	47	82
HighPrice	22.3	159	27	62
HighPrice large adjusted	43.4	175	56	101

[1]EEBI score, forgone net agricultural revenue, and rental payments are computed using the 30 million acres "accepted" by the LTB simulation.

LTB = likely to bid; EEBI = Environmental Benefits Index, without the cost factor; RFS = Renewable Fuel Standard.

Two variants of the *MediumPrice* scenario are examined: one that uses the 2007 CRP rental rates and a second that increases these rental rates across the board.[15] Far fewer acres (28.8 million) are offered in the first variant (which uses 2007 CRP rental rates) than in the *Baseline*, a result that we attribute to the run-up in prices that began in 2007. That is, the gap between a landowner's projected revenue and CRP payments accounts for the fewer acres offered in the *MediumPrice* scenario. The second variant (the *MediumPrice adjusted* scenario) adjusts for this gap by raising CRP rental rates 60 percent, which causes offered acres to rebound to 45 million (table 5).[16]

Now consider our stylized *RFS adjusted* scenario, which accounts for commodity price impacts as ethanol output grows from 6.5 to 15 billion gallons, and which uses the same CRP rental rates as the *MediumPrice adjusted* scenario. Here, crop prices are slightly higher than in the *MediumPrice adjusted* scenario. Since per-acre CRP rental payments are the same in both scenarios, we would expect a slight contraction in offers under the *RFS adjusted* scenario, and in fact offered acres do decline about 2 million (table 5). The average EEBI score decreases by about 2 percent.

To further illuminate the impacts of increasing ethanol production, we track individual data points across the *MediumPrice adjusted* and the *RFS adjusted* scenarios (table 6):

- Of the 30 million accepted acres in both scenarios, about 1.8 million acres differ. That is, there are 28.2 million acres that are accepted under either scenario, and 1.78 million acres that are replaced in the *MediumPrice* adjusted scenario with different acres in the *RFS adjusted scenario.*
- Under the *RFS adjusted* scenario, exiting CRP acreage tends to be more productive land that received higher rent and delivered a higher EEBI. The new acres entering the program (as crop-based ethanol production ramps up) are more expensive (in terms of agricultural revenue and rental payments) than the average retained acre.
- New acres have lower average EEBI scores than either retained acres or exiting acres.[17]

In other words, the additional acres accepted under the *RFS adjusted* scenario are acres that failed to make the cut in the *MediumPrice adjusted* scenario, but become acceptable when higher quality, but more expensive, acres bypass the program due to a higher crop returns.

As ethanol production increases in the *RFS* scenario, CRP acreage declines by 6.5 percent in the Corn Belt and 3.5 percent in the Lake States, while increasing 3.8 percent in the Northern Plains (table 7). Thus, the impacts of higher prices in the *RFS adjusted* scenario are more pronounced in more productive regions.

Choosing CRP Rental Rate Increases

Changes in commodity prices are likely to lead to changes in cropland rental rates. Since the USDA bases its CRP rental rates on county averages of non-irrigated cropland rents, changes in commodity prices will be reflected by changes in the CRP's rental rates.

However, as discussed in Appendix 3, just what these changes will be is uncertain. USDA's methodology to establish CRP rental rates combines surveyed values, such as provided by USDA's National Agricultural Statistics Service (http://quickstats.nass.usda.gov/?source_desc=SURVEY&commodity_ desc=RENT&agg_level_desc=COUNTY), and feedback from county and State shareholders. Thus, it is not straightforward to predict what rental rates would be under different commodity price regimes.

Therefore, rather than attempting to construct detailed rental rate increases (say, that capture the possibility of higher increases in Iowa than in Texas), we use across-the-board proportional increases. While a number of levels were simulated, for brevity we present results from simulations that use a 60-percent across-the-board increase (i.e., the *MediumPrice adjusted*) and a 120-percent across-the-board increase (i.e., the *HighPrice large adjusted*).

The choice of 60 percent (and 120 percent) was predicated on overall plausibility: they yield predicted CRP enrollment that retains most of the program's current environmental characteristics without an implausible increase in program costs. That is: these rental rate increases are meant to represent a possible outcome, they are not meant to be an actual prediction of what will occur.

For comparison purposes, note that (using county averages) actual CRP rental rates increased by 33 percent between 2000 and 2008, and by 20 percent between 2008 and 2010.

Most importantly, although these across-the-board increases may not capture the actual increases that come to pass, they allow us to compare changes across simulations—comparisons that are the main goal of this research.

Table 6. Acres entering and leaving the CRP general signup as crop-based ethanol production increases to 15 billion gallons

	For acres that doFor acres thatFor acres that not change as ethanol production increases[1]	For acres that enter the CRP as ethanol production increases	For acres that leave the CRP as ethanol production increases
Forgone net revenue (avg. $/acre)	47	61	137
Rental payments (avg. $/acre)	81	97	113
EEBI score (avg. score/acre)	178	147	192

[1] This table is based on a comparison of acres offered, and accepted, into the CRP general signup program under the MediumPrice adjusted scenario and the RFS adjusted scenario, using the LTB model. In each scenario, 30 million acres are accepted; 28.2 million acres are accepted in both scenarios, with 1.8 million acres differing. Note: EEBI = Environmental Benefits Index, without the cost factor.

Table 7. Geographic distribution of acres: LTB model predictions of MediumPrice adjusted, RFS adjusted, and HighPrice large adjusted scenarios

				Farm Production Region[2]						
	NE	AP	SE	DE	CB	LS	NP	SP	MT	PA
			Percent of all CRP general signup acres							
MediumPrice adjusted	1.5	3.8	5.5	3.5	10.5	10.1	29.2	11.8	18.6	5.0
RFS adjusted	1.5	3.7	5.3	3.5	9.8	9.8	30.3	11.9	18.7	5.0
Percent change[1]	-1.6	-0.7	-3.0	1.0	-6.5	-3.5	3.8	0.1	0.5	0.4
HighPrice large adjusted	2.1	3.2	4.1	2.8	8.0	8.4	32.5	11.0	21.9	5.9
Percent change[1]	35	-15	-26	-21	-24	-17	11	-7.0	17	18

[1] Percent change in region, compared to the MediumPrice adjusted scenario.
[2] See table 4 for States in each region.
Notes: LTB = likely to bid, RFS = Renewable Fuels Standard.

We model the *HighPrice large adjusted* scenario using a 120-percent increase in CRP rental rates. Results here (in terms of offered acres and EEBI scores) are similar to the *MediumPrice adjusted* scenario (table 5), though average per-acre rental payments increase to $101/acre. When 2007 CRP rental rates are used (that is, rental rates are not adjusted to accommodate high crop prices), less than 30 million acres of land are offered to the program.[18]

In comparing the "transition" of acres from the *MediumPrice adjusted* to the *HighPrice large adjusted* scenario, substantive changes are evident (table 8). Of the 30 million accepted acres, 7 million acres differ between the two scenarios. Unlike with transitions between prior scenarios, forgone revenues are higher (not surprising, given the high crop prices), as are rental payments.[19] The land that enters under the *HighPrice large adjusted* scenario is both of lower agricultural productivity and lower environmental quality than the land that leaves.

In contrast to the comparison of *MediumPrice adjusted* and the *RFS adjusted* scenarios, regional shifts in CRP acres under the *HighPrice large adjusted* are substantial. The Southeast, Delta, Appalachia, Lake States, and Corn Belt regions lose from 15 to 26 percent of their enrolled acres, while acreage in the Northern Plains, Northeast, Mountain, and Pacific regions expands by 10-35 percent (table 7).

An interesting feature of the CRP is that the program's total allowable enrollment is capped, but there is no explicit budget cap. Nonetheless, program dollar outlays are a concern, as evidenced by inclusion of a cost factor in the EBI, so it is interesting to compare the budget/acreage impacts of the different scenarios.

Figures 2 and 3 display results from five scenarios, along with results from actual CRP contracts as of March 2009. Each graph displays the impact of increasing the number of accepted acres (where acres with the highest EBI scores are accepted first). Each graph examines a single variable. Figure 2 shows the total costs of enrolling various acre levels. Figure 3 shows the average environmental component of the EBI score (the EEBI), per acre, at various acre levels.

Table 8. Acres entering and leaving the CRP general signup under summer 2008 commodity

	For acres that do not change as prices and rental rates increase[1]	For acres that enter the CRP as prices and rental rates increase	For acres that leave the CRP as prices and rental rates increase
Forgone net revenue (avg. $/acre)[2]	51	77	153
Rental payments (avg. $/acre)[2]	104	92	129
EEBI score (avg. score/acre)	178	163	179

[1] This table is based on a comparison of acres offered, and accepted, into the CRP general signup program under the MediumPrice adjusted scenario and the HighPrice large adjusted scenario, using the LTB model. In each scenario, 30 million acres are accepted; 23 million acres are accepted in both scenarios, with 7 million acres differing.

[2] Forgone agricultural revenue and rental payments are based on values from the HighPrice large adjusted scenario.

Note: EEBI = Environmental Benefits Index, without the cost factor.

Figure 2 illustrates that the contracts as of 2009 (many of them dating from 1997) are noticeably less costly than all the scenarios (though program costs are somewhat close to the *Baseline*). The *HighPrice large adjusted* scenario costs nearly three times as much as actual contracts when 30 million acres are enrolled. The *MediumPrice adjusted* and *RFS adjusted* scenarios (whose differences are due to projected changes in ethanol production) are quite similar.

What happens to general signup acreage if (say, due to budget pressures) 2009 funding levels were imposed? The 2009 contracts indicate that about $1.1 billion is spent on the "best" 30 million acres (in terms of EBI score). If this $1.1 billion could not be expanded, under the other scenarios CRP acreage would substantially decline. For example, in the *MediumPrice* scenario, only 20 million acres could be enrolled.

And in the *MediumPrice adjusted* and *RFS adjusted* scenarios (where CRP rental rates are increased across the board), only about 15 million acres could be enrolled, with even less should the commodity prices observed in summer 2008 (the *HighPrice large adjusted* scenario prices) be realized.

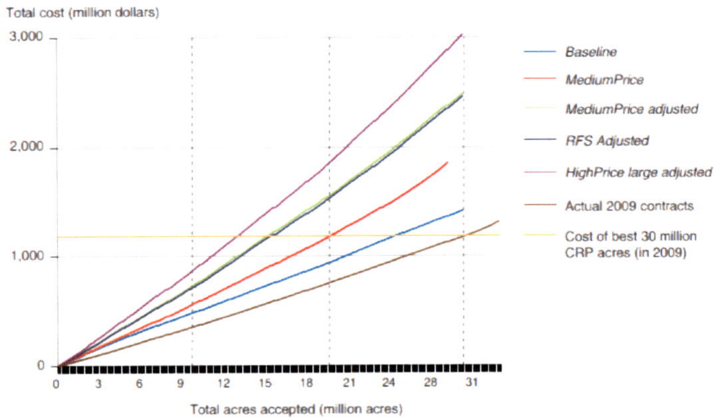

The vertical axis lists the total acquisition cost for a given acreage.

Figure 2. Total cost of CRP acquisitions at different acreages, for several scenarios.

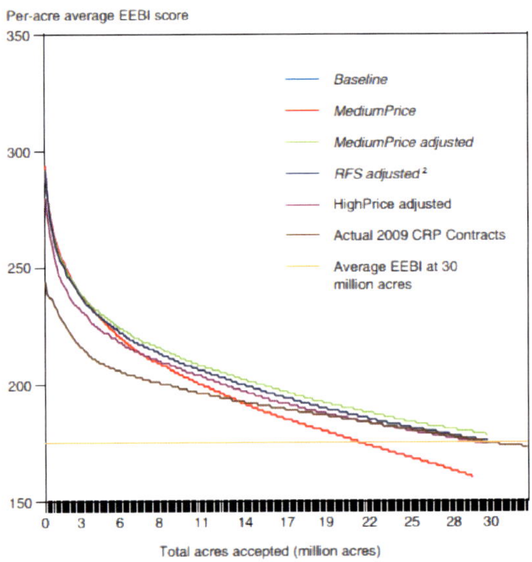

The vertical axis lists the average EEBI across all acres, at a given acreage. For example, the graph for the MediumPrice scenario has a value of 176 at 21 million acres. This means that when the 21 million acres with the best EBI scores (in the MediumPrice scenario) are considered, their average EEBI score is 176.

[1]The EEBI is the Environmental Benefits Index (EBI) without the cost factor.

[2]The RFS (Renewable Fuels Standard) scenario reflects additional price impacts of heightened biofuels production.

Figure 3. Average per-acre EEBI1 score at different CRP acreage levels, for several scenarios.

Conversely, what is the impact of adjusting CRP rental rates? Does offer quality improve? At all acreage levels, the "adjusted" scenarios yield per-acre EEBI scores between those of the 2009 contracts and the 2005 baseline (fig. 3). In fact, the "adjusted" scenarios are quite similar, which confirms that increasing rental payments can help maintain environmental dividends. It is also interesting to note what happens in the *MediumPrice* scenario (where no adjustments to CRP rental rates are made): not only is the 30-million-acre signup goal unobtainable, but the quality of enrolled acres (as measured by average EEBI score) is substantially lower than under current contracts.

INFLUENCE OF PRICES: *OPT-OUT* MODEL

Although the *LTB* model does have appealing features (such as allowing new lands to enter), it ignores a key aspect of the CRP—that there are lands already in the program. And, these lands may be controlled by rural land-owners who embrace the idea of land retirement and may behave differently than the average landowner.

In addition, CRP contracts are for 10 to 15 years, so the notion of "restarting the CRP from scratch" is unrealistic. More likely, owners of lands currently enrolled in the CRP may choose not to re-enroll when their contracts expire. Or, the USDA may enact a policy to let lands more easily opt out of the CRP if more production is deemed necessary.

Therefore, as a complement to the *LTB* model, we formulate an opt-out model that focuses on currently enrolled CRP acreage. This analysis uses yield and crop price information to assign a "crop" to every CRP contract (see appendix 2). This assignation is a prediction of the crop that the land would return to without the CRP. Unlike other analyses (Lubowski et al., 2008), idling land is not an option. The CRP land is assumed to return to some kind of crop production (perhaps with a few years' lag).

The *opt-out* model can indicate the consequences of releasing CRP acreage. For example, to meet increased demands for crops (say, due to a large expansion in crop-based ethanol production), one approach is to release CRP land before its scheduled expiration date—to allow contracted acres to opt out. If such a policy were enacted, what would be the consequences of releasing acreage at different levels? To address this question, we sort contracts that our models predict would be likely to opt out and examine the tradeoffs if more or fewer acres are allowed to opt out.

The *opt-out* model uses contract data as of March 2008.[20] It assigns acreage in all CRP contracts to a crop type and predicts what would happen to program acres if the CRP were to abruptly end. Using this prediction of probable crop choices on current CRP lands, the model then predicts which landowners would opt out of the program if costless exit were permitted. These opt-out predictions are informed by parcel-level estimates (derived from the LTB model) of the correlation between the profitability of the CRP (relative to crop production) and the probability of being enrolled in the CRP.

We first consider scenarios using the prices from the *MediumPrice* scenario. Commodity prices used in these scenarios are higher than those prevalent when CRP contracts were signed. Thus, under the assumption that CRP rental rates do not change, a significant number of acres—9.4 million or 30 percent of general signup acres as of March 2008—would leave the program if given the chance (table 9). Note that this analysis excludes the 4 million acres of continuous signup. By way of comparison, in 2006 CRP offered an extension option to CRP participants whose contracts were set to expire between 2007 and 2010. Approximately 15 percent of the acreage covered by these contracts (about 4 million out of 27 million) opted out.

The *opt-out* model predicts that about half of acres exiting CRP would go into wheat and about a quarter would become either corn or soybeans (table 9). The probability that a given acre will exit the CRP is a function of what crop type the model assigns to it. For example, for acres predicted to "return" to corn and soybeans, the opt-out rate is 32-35 percent, while for sorghum the rate is below 20 percent.

Next we consider the *RFS* scenario (table 10), which captures price increases due to expanding ethanol production from 6.5 billion to 15 billion gallons, without changing CRP rental rates from their 2007 levels. This expansion has a minor impact on CRP enrollment, with about 200,000 additional acres (from 9.4 to 9.63 million acres) choosing to leave the program. The most noticeable change is the 20-percent increase (from 32 to 39 percent) in the opt-out rate for acres assigned to corn, even though the share of exiting acres assigned to corn increases by a smaller amount.

Table 9. Acreage opting out of the general signup CRP by crop: MediumPrice scenario prices, with and without an increase in CRP per-acre rental rates

	Barley	Corn	Cotton	Oats	Sorghum	Soybeans	Wheat	Hay	Acres leaving *(million)*
Crop grown if CRP ended *(%*									
of all current CRP acres)[1]	3	13	3	1	3	12	50	15	
Share of above acres that would opt out of the CRP[2]	35	32	1	4	17	35	40	4	9.4
As above, with 60% increase in soil rental rates (*Medium-*									
Price adjusted scenario) 2	17	11	1	2	11	20	26	2	5.9

[1] The first row is based on a crop prediction: each CRP contract is "assigned" a crop that would be grown if the CRP ceased to exist (see appendix 2 for a description).

[2] The second and third rows use this crop prediction and estimate what share of acres (that are assigned to a given crop) would opt out of the CRP. The second row uses CRP rental rates from the MediumPrice scenario, while the third row uses CRP rental rates used in the MediumPrice adjusted scenario.

Table 10. Acreage opting out of the general signup CRP by crop: RFS scenario prices, with and without an increase in CRP per-acre rental rates

	Barley	Corn	Cotton	Oats	Sorghum	Soybeans	Wheat	Hay	Acres leaving *(million)*
Crop grown if CRP ended									
(% of all current CRP acres)[1]	3	15	3	1	4	11	48	15	
Share of above acres that would opt out of the CRP[2]	35	39	1	4	17	35	40	5	9.6
As above, with with 60% increase in soil rental rates									

Table 10. (Continued)

	Barley	Corn	Cotton	Oats	Sorghum	Soybeans	Wheat	Hay	Acres leaving *(million)*
(RFS adjusted scenario)[2]	17	25	1	2	11	19	27	3	6.3

[1] The first row is based on a crop prediction: each CRP contract is "assigned" a crop that would be grown if the CRP ceased to exist (see appendix 2 for a description).

[2] The second and third rows use this crop prediction and estimate what share of acres (that are assigned to a given crop) would opt out of the CRP. The second row uses CRP rental rates from the RFS scenario, while the third row uses CRP rental rates used in the RFS adjusted scenario.

Note: RFS = Renewable Fuel Standard.

If CRP rental rates are raised 60 percent (the *adjusted* scenarios), the number of acres that opt out is reduced in both the *MediumPrice* and *RFS* scenarios by about 35 percent. Except for corn, this level of reduction is consistent across crops. For corn, the use of higher rental rates causes far fewer acres to opt out in the *MediumPrice* scenario than in the RFS scenario, indicating the allure of biofuels-driven price increases for growing corn.

Under the *HighPrice* scenario, the model predicts that over 11 million acres (about 35 percent of total acres) would opt out. The most noticeable change from other scenarios is for soybeans, with more CRP acres predicted to go into soybeans, and more of these soybean acres predicted to opt out. Under the *HighPrice large adjusted* scenario (where rental rates are increased by 120 percent), the opt-out rate for all crops drops by more than 50 percent, with especially large changes in corn, soybean, and wheat opt-out rates.[21]

Results across scenarios indicate that much of the probable loss of CRP acres is likely due to price changes that have already happened, as indicated by the results under the *MediumPrice* scenarios. Additional changes, as reflected in the *RFS* and the *HighPrice* scenarios, increase opt-out rates somewhat, but much of this could be avoided with higher rental rates, albeit at higher overall program costs.

These findings reinforce the results from the *LTB* models. The impacts of current prices on CRP enrollment are substantial. Further price increases (and CRP effects) due to a large upswing in ethanol output are minor, while CRP impacts with crop prices as high as in summer 2008 are larger.

Table 11. Acreage opting out of the general signup CRP by crop: HighPrice scenario, with and without an increase in CRP per-acre rental rates

	Barley	Corn	Cotton	Oats	Sorghum	Soybeans	Wheat	Hay	Acres leaving (million)
Crop grown if CRP ended (% of all current CRP acres)[1]	3	17	4	1	4	17	48	5	
Share of above acres that would opt out of the CRP2	51	38	0	9	17	43	42	1	11.1
As above, with with 120% increase in soil rental rates (HighPrice large adjusted scenario)[2]	25	20	0	1	9	17	15	1	4.6

[1] The first row is based on a crop prediction: each CRP contract is "assigned" a crop that would be grown if the CRP ceased to exist (see appendix 2 for a description).

[2] The second and third rows use this crop prediction and estimate what share of acres (that are assigned to a given crop) would opt out of the CRP. The second row uses CRP rental rates from the HighPrice scenario, while the third row uses CRP rental rates used in the HighPrice adjusted scenario.

Since a likely goal of releasing CRP lands is to either increase commodity production or reduce program costs, our model allows the most productive land (having the highest rental rates) to be released first. The CRP's Rental Rate is a reasonable measure of productivity, so we sort acres that want to opt out by decreasing rental rate. Note that acres in continuous signup are not released; our analysis indicates that about 300,000 acres in continuous signup would be candidates to opt out.

Table 12. Effects of allowing general signup CRP contacts to opt out: MediumPrice scenario

Cumulative acreage opting out[1]	Rental rate[2]	Annual cost decline[3]	EEBI per acre[4]	Corn share per acre[5]	Corn yield per acre[6]	Wheat share[5]	Wheat yield per acre[6]
Million acres	$/acre	$ millions		Percent	Bushels	Percent	Bushels
1	69	88	203	44	141	14	61
3	39	194	184	28	132	37	50
5	32	265	173	19	131	53	41
8	23	348	163	13	128	65	35
9.4	7	377	161	12	127	68	33

Note: Average per-acre rental rate (SRR) for all CRP contracts is $47. EEBI = Environmental Benefits Index, without the cost factor.

[1] Acres interested in opting out, sorted by descending soil rental rates (SRR). For example, the 2nd row refers to the 3 million "highest SRR" acres that would opt out.

[2] The soil rental rate (SRR) of the n millionth acre. Thus, the 8 millionth acre opting out has a SRR of 23.

[3] Cumulative expenditure (over n million acres). For example, if 8 million acres were allowed to opt out, total CRP expenditures would decline by $348 million/year.

[4] The average EEBI (across n million acres). For example, if 8 million acres were allowed to opt out, their average EEBI score would be 163.

[5] The percent (of n million acres) that go into corn [wheat]. For example, if 3 million acres were allowed to opt out, then 28% [37%] of these acres would go into corn/wheat.

[6] Average corn/wheat yield (across n million acres). For example, if 3 million acres were allowed to opt out, the average corn [wheat] yield (of the acres that go into corn/wheat) would be 132 bushels [50 bushels] per acre.

Tables 12 to 17 show the cumulative impacts of releasing different levels of acres from the CRP. For example, if 3 million acres are released under the *MediumPrice* scenario, the average EEBI of these 3 million acres is an estimated 184. If 5 million acres are released, the average EEBI drops to 173, implying that these additional 2 million acres provide fewer environmental benefits than the prior 3 million acres. In all cases, the first acres to opt out tend to go to corn. And, as more acres opt out, the share of wheat acres increases.

Table 13. Effects of allowing general signup CRP contacts to opt out: *RFS* scenario

Cumulative acreage opting out[1]	Rental rate[2]	Annual cost decline[3]	EEBI per acre[4]	Corn share per acre[5]	Corn yield per acre[6]	Wheat share[5]	Wheat yield per acre[6]
Million acres	$/acre	$ millions		Percent	Bushels	Percent	Bushels
1	71	91	204	52	141	11	60
3	41	202	187	37	131	30	51
5	32	274	175	25	129	48	41
8	24	360	165	18	126	60	35
9.6	7	395	161	15	125	64	32

Notes: Average per-acre rental rate is $47. RFS = Renewable Fuel Standard, EEBI = Environmental Benefits Index, without the cost factor.

[1] Acres interested in opting out, sorted by descending soil rental rates (SRR). For example, the 2nd row refers to the 3 million "highest SRR" acres that would opt out.

[2] The soil rental rate (SRR) of the n millionth acre. Thus, the 8 millionth acre opting out has a SRR of 24.

[3] Cumulative expenditure (over n million acres). For example, if 8 million acres were allowed to opt out, total CRP expenditures would decline by $360 million/year.

[4] The average EEBI (across n million acres). For example, if 8 million acres were allowed to opt out, their average EEBI score would be 165.

[5] The percent (of n million acres) that go into corn [wheat]. For example, if 3 million acres were allowed to opt out, then 37% [30%] of these acres would go into corn [wheat].

[6] Average corn/wheat yield (across n million acres). For example, if 3 million acres were allowed to opt out, the average corn [wheat] yield (of the acres that go into corn/wheat) would be 131 bushels [51 bushels] per acre.

Comparing across price scenarios, as prices increase (from *MediumPrice* to *RFS* to *HighPrice*), the number of opt-out acres increases (although by a relatively small amount in the *MediumPrice* to *RFS* case). Other statistics are less affected, though corn acres increase slightly (at low opt-out acreage) as one moves to higher price scenarios (i.e., from *MediumPrice* to *RFS*).

Comparing across different variants of each scenario, an increase in CRP rental rates reduces the number of opt-out acres, at higher program cost. In terms of proportional change, the *HighPrice large adjusted* scenario shows the most dramatic within-scenario reduction. Interestingly, across all three pairs of scenarios the cost savings (of the opted-out acres) are roughly the same: larger across-the-board increases in CRP rental rates means fewer acres opt out, but each opted-out acre has a higher average SRR. Increasing the SRR multiplier also seems to reduce the corn share (at a given level of opt-out acres). For example, in the *RFS adjusted scenario*, at 2 million opted-out acres, the corn proportion is 26 percent; in the *RFS* (without a higher SRR) scenario, it is 37 percent. Again, corn growers may be more sensitive to changes in relative net returns in a volatile price environment.

Table 14. Effects of allowing general signup CRP contacts to opt out: HighPrice scenario

Cumulative acreage opting out[1]	Rental rate[2]	Annual cost decline[3]	EEBI per acre[4]	Corn share per acre[5]	Corn yield per acre[6]	Wheat share[5]	Wheat yield per acre[6]
Million acres	$/acre	$ million		Percent	Bushels	Percent	Bushels
1	80	100	209	56	151	03	65
3	52	231	195	39	138	15	60
5	36	316	183	28	135	32	47
8	28	412	171	19	133	48	38
11.1	7	486	164	15	131	58	33

Notes: Average per-acre rental rate is $47. EEBI = Environmental Benefits Index, without the cost factor.

[1] Acres interested in opting out, sorted by descending soil rental rates (SRR). For example, the 2nd row refers to the 3 million "highest SRR" acres that would opt out.

[2] The soil rental rate (SRR) of the n millionth acre. Thus, the 8 millionth acre opting out has a SRR of 28.

[3] Cumulative expenditure (over n million acres). For example, if 8 million acres were allowed to opt out, total CRP expenditures would decline by $412 million/year.

[4] The average EEBI (across n million acres). For example, if 8 million acres were allowed to opt out, their average EEBI score would be 171.

[5] The percent (of n million acres) that go into corn [wheat]. For example, if 3 million acres were allowed to opt out, then 39% [15%] of these acres would go into corn/wheat.

[6] Average corn/wheat yield (across n million acres). For example, if 3 million acres were allowed to opt out, the average corn [wheat] yield (of the acres that go into corn/wheat) would be 138 bushels [60 bushels] per acre.

Table 15. Effects of allowing general signup CRP contacts to opt out: MediumPrice adjusted scenario

Cumulative acreage opting out1	Rental rate[2]	Annual cost decline[3]	EEBI per acre[4]	Corn share per acre[5]	Corn yield per acre[6]	Wheat share[5]	Wheat yield per acre[6]
Million acres	$/acre	$ millions		Percent	Bushels	Percent	Bushels
1	82	115	192	32	129	28	57
3	50	237	171	15	128	59	40
4	44	285	166	13	127	64	36
5.9	11	358	158	09	125	72	32

Notes: Average per-acre rental rate is $78. EEBI = Environmental Benefits Index, without the cost factor.

[1] Acres interested in opting out, sorted by descending soil rental rates (SRR). For example, the 2nd row refers to the 3 million "highest SRR" acres that would opt out.

[2] The soil rental rate (SRR) of the n millionth acre. Thus, the 4 millionth acre opting out has a SRR of 44.

[3] Cumulative expenditure (over n million acres). For example, if 4 million acres were allowed to opt out, total CRP expenditures would decline by $285 million/year.

[4] The average EEBI (across n million acres). For example, if 4 million acres were allowed to opt out, their average EEBI score would be 166.

[5] The percent (of n million acres) that go into corn [wheat]. For example, if 3 million acres were allowed to opt out, then 15% [59%] of these acres would go into corn/wheat.

[6] Average corn/wheat yield (across n million acres). For example, if 3 million acres were allowed to opt out, the average corn [wheat] yield (of the acres that go into corn/wheat) would be 128 bushels [40 bushels] per acre.

Table 16. Effects of allowing general signup CRP contacts to opt out: RFS adjusted scenario

Cumulative acreage opting out[1]	Rental rate[2]	Annual cost decline[3]	EEBI per acre[4]	Corn share per acre[5]	Corn yield per acre[6]	Wheat share[5]	Wheat yieldper acre[6]
Million acres	$/acre	$ millions		Percent	Bushels	Percent	Bushels
1	93	125	197	50	132	18	58
3	52	257	174	26	128	50	41
5	40	350	164	18	126	62	35
6.3	11	395	160	15	124	67	32

Notes: Average per-acre rental rate is $78. RFS = Renewable Fuel Standard; EEBI = Environmental Benefits Index, without the cost factor.

[1] Acres interested in opting out, sorted by descending soil rental rates (SRR). For example, the 2nd row refers to the 3 million "highest SRR" acres that would opt out.

[2] The soil rental rate (SRR) of the n millionth acre. Thus, the 5 millionth acre opting out has a SRR of 40.

[3] Cumulative expenditure (over n million acres). For example, if 5 million acres were allowed to opt out, total CRP expenditures would decline by $350 million/year.

[4] The average EEBI (across n million acres). For example, if 8 million acres were allowed to opt out, their average EEBI score would be 164.

[5] The percent (of n million acres) that go into corn [wheat]. For example, if 3 million acres were allowed to opt out, then 26% [50%] of these acres would go into corn/wheat.

[6] Average corn/wheat yield (across n million acres). For example, if 3 million acres were allowed to opt out, the average corn [wheat] yield (of the acres that go into corn/wheat) would be 132 bushels [41 bushels] per acre.

Table 17. Effects of allowing general signup CRP contacts to opt out HighPrice large adjusted scenario

Cumulative acreage opting out[1]	Rental rate[2]	Annual cost decline[3]	EEBI per acre[4]	Corn share per acre[5]	Corn yield per acre[6]	Wheat share[5]	Wheat yield per acre[6]
Million acres	$/acre	$ millions		Percent	Bushels	Percent	Bushels
1	131	179	198	48	139	9	61
3	65	352	174	25	132	39	39
5	52	411	167	21	131	47	34
4.6	15	441	164	18	130	52	32

Notes: Average per-acre rental rate is $103. EEBI = Environmental Benefits Index, without the cost factor.

[1] Acres interested in opting out, sorted by descending soil rental rates (SRR). For example, the 2nd row refers to the 3 million "highest SRR" acres that would opt out.

[2] The soil rental rate (SRR) of the n millionth acre. Thus, the 5 millionth acre opting out has a SRR of 52.

[3] Cumulative expenditure (over n million acres). For example, if 5 million acres were allowed to opt out, total CRP expenditures would decline by $411 million/year.

[4] The average EEBI (across n million acres). For example, if 5 million acres were allowed to opt out, their average EEBI score would be 167.

[5] The percent (of n million acres) that go into corn [wheat]. For example, if 3 million acres were allowed to opt out, then 25% [39%] of these acres would go into corn/wheat.

[6] Average corn/wheat yield (across n million acres). For example, if 3 million acres were allowed to opt out, the average corn [wheat] yield (of the acres that go into corn/wheat) would be 132 bushels [39 bushels] per acre.

HOW MIGHT A CO_2 OFFSET PROGRAM INFLUENCE CRP PARTICIPATION?

Even without commodity prices higher than those used in the *MediumPrice* scenario, our results indicate that a substantial increase in program expenditures (for example, from $48 to $86 per acre) would be required to maintain the CRP in something close to its present configuration. Other funding mechanisms could reduce costs—we consider payments for carbon sequestration as an example.

In particular, land retirement (along with suitable land cover and management) can lead to carbon sequestration (Pineiro et al., 2009). Conceivably, lands that enroll in the CRP could generate GHG offsets that could be sold (by the landowner). While issues of leakage and additionality may complicate implementation (Baker and Galik, 2009), let us assume that landowners are allowed to sell the carbon sequestration of their retired acres, as measured by the yearly difference in carbon storage when in a conserving practice versus carbon storage (or release) in a continuous crop.[22]

The *LTB* model is used, starting with prices and rental rates from the *MediumPrice* scenario. Sequestration rates by farm production region, which measure the additional carbon that conversion from a continuous crop to CRP would entail (table 18), are assigned to all data points in the *LTB* model. This sequestration is treated as a marketable carbon offset. Thus, a CRP acre will receive both a rental payment from USDA and a carbon offset payment from whoever purchases it.[23]

Table 18. CRP acreage characteristics under the MediumPrice scenario, with an active carbon market

	Offered acres	Average EEBI score	GHG offset revenue	Forgone agricultural revenue	Rental payments
$25/mt CO_2 price, 1.0	Million	Average/acre		Average $/acre	
x soil rental rate	37.1	181	16	42	58
$30/mt CO2 price, 1.3 x soil rental rate	46.9	191	19	53	70
$50/mt CO2 price, 1.0 x soil rental rate	45.5	190	31.6	51	55
MediumPrice adjusted	44.7	179	..	48	83

Note: Columns 2-5 are based on the "best" 30 million acres that the model accepts into the CRP.

The increased carbon sequestration rate, compared to continuous agriculture, per acre of CRP are from Eve et al. (2002); with values—in metric tons (mt) per acre—that range between 0.29 (for the Mountain farm production region) to 0.74 (for the Delta farm production region).

EEBI = Environmental Benefits Index, without the cost factor. GHG = greenhouse gas.

1) A CO_2 price of $25/ton, with no increase in rental rates;
2) A CO_2 price of $30/ton, with a 30-percent increase in rental rates; and
3) A CO_2 price of $50/ton, with no increase in rental rates.
4) If CO_2 prices were as high as $50/ton, the CRP would be largely unaffected by higher commodity prices, even without an increase in rental rates (table 18). At $30/ton (along with a 30-percent increase in rental rates), the *LTB* predicts environmental benefits similar to the $50/ton scenario, but with a 27-percent increase in program costs (from $55 to $70 per acre). Total revenue per acre (rental rate plus offset revenue) in both these scenarios is $87-$89/acre, which is near the $83 rental value of the *MediumPrice adjusted* scenario.

Three scenarios are examined. The first represents a consensus price for carbon offsets, and the latter two are selected to achieve results, in terms of EEBI scores and offered acres, that are similar to the current CRP.

CONCLUSION

In this study, we examine how high commodity prices may affect the Conservation Reserve Program. As a voluntary program, landowners choose whether to offer their land to the program or to engage in agricultural production. Higher crop prices, as observed in summer 2008 or subsequent to mandated increases in crop-based ethanol production, are likely to sway some landowners in favor of agricultural production over conservation.

The impacts of a change in the economic environment depend on a number of factors. Of course, the magnitude of price changes matters. So does the reaction of the USDA program administrators, such as by increasing per-acre CRP rental rates as per-acre agricultural profits rise. The future goals of the program also matter, as CRP size is diminished and as the more closely targeted continuous signup portion of the program grows in importance. Also, commodity price changes influence not only the amount but the quality of lands offered: their agricultural productivity, environmental benefits, and geographic location.

We consider these issues using two modeling strategies. The *likely-to-bid (LTB)* model simulates what the CRP would look like if it were to start from scratch. The *opt-out* model takes current CRP contracts as a given, then simulates which enrollees would opt out of the program if given the opportunity. Both strategies have their strengths and weaknesses, and together offer complementary perspectives on the impacts of price changes.

With a mandated increase in biofuels production, one scenario (*RFS*) simulates the impact on crop prices due to higher crop-based ethanol output. Another scenario (*HighPrice*) incorporates the high prices observed in summer 2008, should crop prices settle at this higher level due to increased demand or other factors.

One key finding is that increased ethanol output would have small additional impacts on the CRP. Using the *LTB* model, offered acres drop by about 2 million acres (5 percent), and environmental benefits (as measured by the environmental component of EBI scores) decrease by about 2 percent. The *opt-out* model tells a similar story, with about 200,000 additional acres leaving the program.

Overall, these results are commensurate with the relatively small change in prices predicted by the REAP model (Biomass Research and Development Initiative, 2008) for this increase in ethanol production. However, these impacts are on top of impacts due to the high commodity prices of the last 3 years. We find that restarting the CRP, under prices and CRP rental rates similar to those currently observed (*MediumPrice* scenario), would lead to significant drops in offered acres and EBI scores and an increase in program costs. We also find that, if contracts were allowed to opt out (without any penalty), almost one-third might do so under the *MediumPrice* scenario.

Thus, in a new era of elevated crop prices—even without an increase in ethanol production—maintaining the CRP as currently configured will require an increase in rental payments. Given the established interest in the program, it might be possible to meet enrollment goals (say, 30 million acres of general signup) with moderate increases in CRP rental rates. But this will likely mean accepting lower average EBI scores, as landowners with profitable but environmentally sensitive lands choose not to offer their land to the program.

With a goal of maintaining the current attributes of retired acres, the *LTB* model finds that in the *MediumPrice* scenario, program costs will almost double (from about $45/acre to about $85/acre). In a similar vein, the opt-out model indicates that this level of rental rate increase reduces the number of acres that choose to opt out by about one-third.

Comparison of enrollment schedules demonstrates that if expenditures are capped at current levels and CRP rental rates are held constant, acreage enrollment will decline precipitously. Increasing the rental rate can offset this decline and lead to enrollment of acreage with environmental benefits comparable to current contracts, but at substantially higher program expenditures. In all scenarios, the *opt-out* model suggests that opt-out acres are most likely to end up in corn. However, as the level of acres allowed to opt out increases, wheat dominates.

If carbon offsets could be sold by CRP enrollees, and if additionality concerns are ignored, the impacts of increased prices on CRP enrollment costs could be substantially reduced. For example, at a price of $50/ton of sequestered carbon dioxide, CRP acreage would be mostly retained, with a minor (about 16-percent) increase in cost.

In summary, as crop prices rise, the composition and environmental attributes of CRP acreage can be maintained so long as rental rates are updated to reflect current crop prices. Such an updating will not be cheap, and within 10 years (as existing contracts expire) may almost double program costs. To the extent such an increase in rental rates does not occur, the program is likely to see fewer acres offered, with a commensurate decrease in the EBI scores of accepted acres.

APPENDIX 1. *LIKELY-TO-BID* MODEL: METHODOLOGY

The *likely-to-bid* (LTB) model, developed jointly by USDA's Economic Research Service (ERS) and Farm Service Agency (FSA), uses data from a number of sources. Field-specific data on land use and land quality are from the 1997 National Resources Inventory (NRI). NRI data are collected and maintained by USDA's Natural Resources Conservation Service (NRCS) for more than 800,000 points of land throughout the contiguous United States.

For each NRI data point, the LTB model determines CRP eligibility using point-specific information on cropping history, soil erodibility, location (within a conservation priority area), and other factors.

For these eligible points, the LTB assigns or calculates:

(1) Farm Service Agency (FSA) per-acre soil rental rate (SRR),
(2) an estimate of net returns to agricultural production, and
(3) an estimate of the Environmental Benefits Index (EBI) score.

The SRR is computed using average non-irrigated cropland rental rates for the county (obtained from USDA's National Agricultural Statistics Service (NASS)), adjusted for soil quality using the Soil Rating for Plant Growth (SRPG) obtained from NRCS. Using NRI cropping history data, agricultural net returns (NRET) are estimated using State-level information on crop prices (NASS), production costs (ERS), and other information (such as hay prices).

The ratio of soil rental rate over agricultural net returns (SRR/NRET) is used to determine whether the landowner would be interested in CRP enrollment. If the ratio exceeds a region-specific threshold value, the land-owner is assumed to be "interested" in the program—he or she is "likely to bid." For most of the country, a threshold value of 0.8 is used to reflect the fact that CRP returns are risk free. That is, the model assumes that risk-averse producers would be willing to accept a lower return in exchange for the reduction in risk realized through CRP enrollment. In some regions, however, land has significant "option" value—such as the value of converting lands to housing or commercial development. Land cannot be sold or converted to another use while a CRP contract is in force. To account for lost option value, threshold values are set at 1.3 in the Northeast and 1.2 in the Appalachian States.

To determine the EBI score, a prediction of the value for each of the EBI factors is required. For most factors, NRI variables can be used to compute factor scores. For example, the EBI's soil erodibility factor can be assigned using the NRI's water (and wind) erodibility variables. Similarly, the NRI sample point's location can be used to assign a value to the water quality zone factor.

Since it depends on specific practices applied, the EBI wildlife factor cannot be exogenously assigned. Therefore, the LTB model assigns conservation practices to each NRI point using information on eligible practices and geography, such as whether the point is located in a State conservation priority area. Practices are chosen to replicate the observed distribution of practices in the CRP as of 2005. Once practices are assigned, the EBI wildlife factor scores can be calculated.

Threshold Values in the *LTB* Model

The LTB model treats NRI points as parcels of land whose landowners may be interested in the CRP. The LTB assumes that its prediction of net agricultural returns more accurately measures landowner profits than the soil rental rate (SRR) assigned to a parcel. Given this assumption, the LTB finds parcels where the net agricultural returns are low, relative to what the CRP would offer in rental payments, and assumes that owners of these parcels will take advantage of this discrepancy and offer the parcel to the program. If the SRR were a perfect predictor of net agricultural returns, then the LTB would predict that all (or none) of the eligible acres would be offered to the program.

Several factors can yield low net agricultural returns. For example, using 2005 data, the model finds a weak correlation (0.21) between low soil quality (as measured by the SRPG) and willingness to enroll. This implies that the SRR does not fully capture the effects of soil quality when it is lower than the county average.

The use of the 0.8 (and 1.2 and 1.3) thresholds is based on historical use of the LTB (to help determine an EBI cutoff in several general signups). These values, while not entirely arbitrary, are not derived by a formal modeling process. That is, the LTB model does not reference actual CRP contracts.

Calibrating the LTB model with actual contract and offer data, with a goal of deriving point-specific thresholds, would be a real improvement. However, a number of calibration methods were attempted, none of which yielded clearly superior results.[1]

Thus, the ad hoc (but honored by prior use) thresholds (0.8 for most of the country) are used. While this is not an ideal approach,[2] it should not seriously bias the overall results obtained from comparing simulations. (Unless landowner behavior changes dramatically as commodity prices change, differences across simulations should be relatively robust to the exact choice of thresholds.)

1A serious difficulty is presented by the existence of county caps (by law, no more than 25 percent of a county's cropland may be enrolled in the CRP). These caps are constraining; for example, LTB discards almost a third of "interested" parcels due to county cap reasons.

2 Solutions to these and other modeling issues should become more precise as more recent data (such as 2007 NRI data) become available.

Once the environmental factors (or the EEBI) and the cost factor have been assigned, a full EBI score is computed for each NRI point. The NRI points are then sorted by decreasing EBI score so that the "first accepted acres" are from sample points with the highest EBI scores. A high score could be due to a low SRR (a large cost component score) or from high environmental values (high EEBI score). Moreover, a point of land with high agricultural productivity (a high SRR) will have a high EBI score only if it has high environmental attributes (a high EEBI). This implies that "accepted" points with high SRRs will also have high EEBIs.

The *LTB* model can be used for scenario analysis by varying what EBI score to use as a cutoff level, and by changing variables that affect allocation of NRI points to the simulated CRP. These variations include estimates of SRR values, crop prices, crop yields, crop variable costs, and the EBI factor weights. *LTB* simulations can be compared statistically, say by examining average EBI scores of accepted acres, at a national or regional level. By combining simulations, one can also simulate land-use changes, such as the distribution of land (NRI points) moving into or out of the CRP.

The *LTB* model is run once per scenario. Additional robustness could be achieved by running the model several times per scenario and averaging the results, with each run distinguished solely by different draws of the random numbers used at several steps (i.e., to assign a practice to a NRI data point).

However, given the intrinsic size of the NRI (800,000 points), a single run should capture the "law of large numbers" benefits that would be provided by multiple runs on a smaller dataset. Hence, the *LTB* is run only once per simulation. Note that to focus on changes of interest (say, increasing commodity prices), where random values are used (i.e., assigning wildlife factor points), the values used (for a given point) are the same across simulations.

Last, the cost factor component of the EBI is based on the SRR assigned to the point of land sampled by the NRI. We assume that "landowners" of these "representative parcels" would ask for the maximum allowed rent (the SRR). While landowners can ask for less than this maximum (and thereby increase the size of the cost factor), in actual practice most landowners do not reduce their offer. Over the last several signups, the average bid in CRP offers (as a share of the parcel's SRR) has been over 95 percent.

APPENDIX 2: OPT-OUT MODEL: METHODOLOGY

The CRP is in place. Evidence suggests that many enrollees have been happy with the program. For example, in 2007 when enrollees were offered the chance to extend their contracts (from 2 to 5 years), with some offered immediate re-enrollment (for 10 years), about 85 percent of all enrollees accepted. While these offers occurred prior to the run-up in crop prices, the rate of acceptance does suggest that the attitudes of current CRP contract holders might be different than those of non-contract holders.

Therefore, using general models to measure changes in the CRP (as prices change) may miss important details on probable participants, details that may be captured by working with actual CRP contracts.

However, CRP contract data lack crucial information. In particular, no variable indicates what contract holders would do with their land if it were not in the CRP.

To analyze the impacts of changes in commodity prices, in the *opt-out* model we compare future agricultural profits to current CRP rents. As agricultural profits increase, a greater share of land currently in the CRP may revert back to crops. So, computing agricultural profits for the non-CRP alternatives (for every contract) is necessary, requiring a guess as to which crop would be grown on the land in the absence of the program.

To formulate this guess, where possible, we use the REAP model to guide us. In particular, we posit some scenario (say, changed prices), modify the

REAP model accordingly, and use the resulting REAP predictions to help assign CRP contracts to crop choices.

REAP is a partial-equilibrium model that determines U.S. regional crop and livestock production levels and national prices for 10 major crops and several livestock and agricultural product categories (Johansson et al., 2007). Import, export, and storage flows are included.

The following describes this "REAP-based" assignment of contracts to crops:

1. For a given scenario (say, the RFS-mandated increase in biofuels production), REAP is used to predict the distribution of crop shares (for all cropland acres).
   a) For each of several crop types, a predicted "cropland share" is computed. This is the share of cropland acres (in a REAP region) that are in this crop type.
   b) A separate schedule of cropland shares is computed for each of the 46 different REAP regions.
   c) The REAP scenario also provides estimates of the prices and variable costs for each crop type (on a national basis).
   d) The crop types are barley, corn, cotton, oats, sorghum, soybeans, wheat, and hay.
2. County-level yield information is derived from NASS statistics (averaged between 2000 and 2005).
3. Relative productivity (RP) information is derived for each contract. This measures how much more (or less) productive this contract is than its county average (for the year of the contract).
   RP is defined as: ContractSRR/AverageSRR_inCountyYear Example:
   Contract JXT13 in county 19037 has an SRR of 135
   The average SRR in county 19037 is 144
   Then: RP = 135/144 =0.87
   Thus, if the average soybean yield in 19037 is 50 bu/acre
   then contract JXT13's predicted soybean yield is 0.87 * 50 = 43.5.
4. To account for the possibility that acres in the CRP tend to be systematically different from observationally equivalent acres not in the CRP, State-specific adjustment factors are computed for each crop type (see box, "Computing an Adjustment Factor").
5. Compute an adjusted crop share, using the REAP share (from step 1) and the State-specific adjustment factor (from step 4) for each crop

type. These are normalized to sum to 1.0. Note that all points in a REAP region will have the same adjusted crop share values.

6. Each NRI point is located within a REAP region and randomly assigned a crop, such that the assigned distribution of crops equals the predicted REAP-region crop shares.
7. Using price and variable cost information associated with the REAP scenario, yield information (from NASS statistics) for this crop (chosen in step 6) and a contract's RP (from step 3), compute net agricultural returns for this randomly assigned "crop."
8. Compare the net agricultural returns from step 7 to the contract's scenario-specific SRR. If the net agricultural returns are sufficiently high, this contract is assumed to want to "opt out" of the CRP (or not re-enroll) and will go into the randomly assigned crop (assigned in step 6). The determination of "sufficiently high" is based on a State-specific probability of enrollment schedule (see box "Computing a Probability of Enrollment Schedule").

Computing an Adjustment Factor

It is possible that acres actually in the CRP are systematically different from identical (in terms of land characteristics) acres that are not in the CRP. To account for this possibility, NRI data are used to compute differences in cropland shares between CRP and non-CRP lands. These differences are then used to compute adjustment factors.

For each State, an adjustment factor for each crop type is computed using the following algorithm:

1. Divide the 1997 NRI into CRP points and cropland points (discard points not in one of these categories).
2. For points that are in the CRP in 1997, look up the land use in 1982 (pre-CRP), and compute (State-specific) crop shares for each type t (t=corn, wheat, etc.):

   $inCRPShare_t$ = 1997 CRP acres in crop t in 1982 / total 1997 CRP acres
3. For the 1997 non-CRP acres, do the same (compute crop shares) $nonCRPShare_t$ = 1997 nonCRP acres in crop t in 1982 / total 1997 nonCRP acres

4. For each crop type (t), compute an adjustment factor: $crpAdjust_t$ = $inCRPshare_t$ / $nonCRPshare_t$

For example, if a State's CRP acres had a high share of corn in 1982 (if $inCRPshare_{corn}$ > nonCRPsh$_{arecorn)}$, then one might expect that CRP acres in this State are more likely to go into corn than the State's typical cropland acre.

The $crpAdjust_t$ adjustment reflects systematic differences between CRP and non-CRP lands. Values greater than 1.0 indicate crops that the CRP land was more likely to have been from (based on State-level trends), and hence more likely to return to.

The construction of this adjustment factor assumes that 1982 land uses are reflective of what might happen 25 years later. Since agronomic and economic factors have changed (i.e., average corn yield has increased from about 110 to 150 bu/acre), one may question the validity of this measure.

However, this measure considers the relative relationship between CRP and non-CRP land, in terms of overall agricultural productivity. If this has not changed significantly, this adjustment factor is likely to be valid. Such constancy, between CRP vs. non-CRP, is much more plausible.

Computing a Probability-of-Enrollment Schedule

How can one determine whether a contract will stay in the CRP should its agricultural profits, or soil rental rate, change? While a generic rule of thumb could be used, such as the "0.8" factor used in the LTB model, a more data-driven approach is available. This approach uses farm production region-specific probability-of-enrollment schedules that are calculated as follows:

Step 1. For each CRP contract, 2005 information is used to generate baseline information.

a) Cropland share information (averaged across farm production regions (FPR) using NASS data from 2000 to 2005) is used to assign each CRP contract to a crop. This is similar to the REAP method described in the *Opt-out* methodology, with actual crop shares used rather than REAP predictions.
b) 2005 commodity price and variable-cost information are then used to compute a baseline netAgReturnContract (net agricultural profit) for this contract.

Basically, this is a guess as to what the profitability of this contracted land would have been in 2005, if the contract had not been in the CRP.

Step 2. Compute probability-of-enrollment schedules. A separate schedule is computed for each FPR.

a) Using the predictions from an LTB model run with 2005 price information, retain all NRI points deemed eligible for the CRP. For these "eligible-for-CRP points," extract the point's Soil Rental Rate (SRRNRI) and net agricultural return (netAgReturnNRI). Note that both of these are computed by the LTB model.
b) Compute an SRR profitability ratio for each eligible NRI point: SRRprofit= SRRNRI / netAgReturnNRI.
c) Recode SRRprofit to a class (from 1 to 50): SRRprofitClass= int (SRRprofit/0.1).

Note that an SRRprofitClass of "1" means "SRR is much less than agricultural profit," while a value of 50 means "SRR is much higher than agricultural profit."

d) For all NRI points in each of the n=1..10 farm production regions, and for c=1..50 (c is a SrrProfitClass computed in step 2c), compute: $probEnroll_n(c)$ = CRP acres in class c in FPR n / all cropland acres in class in FPR n

To compute the numerator, we use an NRI variable that indicates whether a point is enrolled in the CRP.

Each probEnrolln is smoothed (across c=1..50), and forced to be non-decreasing.[1]

This yields 10 FPR specific schedules that relate the relative profitability of the CRP (the SRRprofitClass) to a probability of being in the CRP (the probEnroll).

[1]That is, probEnrolln is adjusted so that the predicted probability of enrolling never decreases as the SRRprofitClass increases. In some regions, this is not always the case. Thus, this adjustment represents a fairly strong assumption: that (*ceteris paribus*) increasing the relative profitability of CRP enrollment will never lead to a reduction in enrollment probability.

Step 3. For each contract, compute baseProbEnroll.

a) Using the contract's actual SRR, and the contract's 2005 netAgReturnContract (derived in step 1b), compute cc: the 2005 SRRprofitClass for this contract.
b) Using the appropriate (FPR-specific) schedule computed in step 2e, compute baseProbEnroll=probEnroll_n (cc).

This baseProbEnroll is used in the next step to help construct a *retain* probability for a contract, given new conditions (say, new prices).

Step 4. Given a new scenario, for every contract:

a) Predict a new SRR and a new netAgReturnContract (using the crop assigned to this contract in step 1a).
b) Using these, calculate a new SRRprofitClass.
c) Using this new SRRprofitClass and the appropriate (FPR-specific) probEnrolln schedule, look up a scenarioProbEnroll.
d) Create a normalized probability that a contract stays in the CRP: ProbRetain= scenarioProbEnroll/ baseProbEnroll.
e) If probRetain>1, set it equal to 1.0; otherwise, use probRetain to randomly decide if this contract stays in the CRP.

Basically, probRetain is the probability that this contract will stay in the CRP under this new scenario.

The underlying notion is that contracts currently in the program may be unusual and not necessarily typical of all lands. For example,

- Contract XX, which is in FPR 1, has an srrProfitClass of 18.
- In FPR 1, probEnroll1 (18) = 20%.
- Since this landowner is enrolled, and since only 20% of "similar" landowners in FPR 1 are enrolled: then this landowner is more interested in enrolling than (at least) 80% of other similar landowners.
- Assume that under a new scenario the contract's srrProfitClass equals 15, and probEnroll1(15)= 15%; then probRetain=15/20 = 0.75.

This implies that 75 percent of the time this landowner will stay in the program. In other words, the 15 percent to 20 percent most "interested" landowners will now want to drop out of the program (which is a quarter of contracts that *ceteris paribus* were in profit class 18). Given our lack of further information on this contract, that means a quarter of the time the contract will want out (or, three-quarters of the time the contract will not want out).

Opt-Out Model: Modifications under the HighPrice Scenario

The REAP-based model provides a fairly rigorous means of predicting what crops CRP contracts would go into if the program were to end. Because REAP is an equilibrium model that employs average yields and other market factors, it is unable to predict the potential distribution of crops when prices fall well outside the range of baseline prices driven by factors that are not included in the model. In particular, we could not tweak the REAP model to generate the summer 2008 prices (that were used in the *HighPrice* scenarios). Thus, for the *HighPrice* scenario, the above methodology was modified.

The modification to the above (REAP-based) methodology occurs in the computation of crop shares for each contract (step 1 above). A "weighted" combination of observed county-specific crop shares, and crop profitability, is used to predict these crop shares.

(A) Compute current land shares (LS).

a. NASS data from 2000 to 2006 are used to predict State-specific crop acreage shares for each of the major crops. Specifically, land shares are computed by dividing the statewide acreage in each crop by total cropland (and averaging over the several years of data).
b. As described in step 4 above, a "CRP adjustment factor" is applied to these crop acreage shares.
c. The shares are normalized to sum to 1.0

(B) Generate a normalized profitability metric (PT)

a. As described in steps 2 and 3 above, county-specific yield information and contract-specific RP are used to compute contract-specific yields for each of several crops.

b. Price and cost information (for the scenario of interest) is combined with these yields to predict crop-specific profitability (for each contract).
c. The three crops with the highest predicted profit are retained, and a normalized ratio of "profit rate" is computed for each of these crops.

Example:

Assume that contract UK512 has the following net returns (per acre):

Corn: 50
Soybeans: 30 Sorghum: 10 Hay: 15
Wheat: 100

The top three are corn, soybeans, and wheat. The sum of profits is used as a normalizer: 50+20+100=180.

The "normalized profit ratio" would be:

Corn:	50/180 =28%
Soybeans:	30/180 =17%
Wheat:	100/180=55%

This normalized profit ratio is an ad hoc metric, a proxy for the probability that a given crop would be the most profitable choice. An order statistic would generate rankings based on the variability of net returns for each crop on each point. Unfortunately, data of the requisite scope and detail to compute such theoretically superior models are not readily available.

(C) Combine land shares and normalized profitability

To allow both profitability and current land share to influence probabilities, a "weighted" combination is used:

$$prob_i = {LS_i * PT_i} \Big/ {\sum_m LS_m * PT_m}$$

where:

Probi = probability that contract goes into crop type i
LSi = land share in crop type i, this county (from step A);
PTi = "normalized profit" share in crop type i, this county (from step B);
m = the sum is over all eight crops.

This crop assignation probability is then used in the same way as the REAP assignation—a crop type is assigned to each contract by randomly drawing a number and comparing it to the crop assignation probabilities, profits are estimated under the new price scenario, and this profit is used to generate a probability of enrollment.

APPENDIX 3: THE CRP'S SOIL RENTAL RATE

CRP rental rates are based on a Soil Rental Rate (SRR) that is assigned to every parcel offered to the program. The SRR is the maximum amount the landowner can request. In practice, most landowners request a payment equal to, or close to, the assigned SRR.

The SRR is constructed using two components:

1) The *average rental rate for non-irrigated cropland for the county* the offered land is located in. For most of the history of the CRP, the average rental rate was derived from the USDA's Land Value Survey. Starting in 2008, a special NASS survey is used to compute county average rental rates for use in the CRP. Since rental rates can fluctuate, a moving average (over several prior years) is used to compute the average cropland rental rate for use in SRR computations.
2) *A soil-specific adjustment factor.* This is used to increase (or decrease) the county average rental rate by as much as 50 percent. These adjustments are based on measures of crop productivity that are specific to soil types, and are normalized by the average soil productivity across the county. Thus, acres on soils whose crop productivity is greater than the county average will have an SRR greater than the county average cropland rental rate.

Commodity prices do not have a direct impact on the SRR. Rather, changes in commodity prices (or other factors, such as input prices) will affect the profitability of farming. If land markets are competitive, a portion of these changes should be transmitted to rental markets for agricultural lands. The size and speed of this transmission depends on the transience of price changes and the market power of renters versus landowners (Parks et al., 2003).

REFERENCES

Abbott, Philip C., Christopher Hurt, and Wallace Tyner. 2008. "What's Driving Food Prices?" *Farm Foundation Issue Report*, July.

Associated Press. 2008. "Higher Prices Could Pressure CRP Program," *Associated Press Newswires*, 5 April.

Babcock, B.A. 2008a. "When Will the Bubble Burst?" *Iowa Ag Review* 14(1). http://www.card.iastate.edu/iowa_ag_review/winter_08/article1.aspx

Babcock, B.A. 2008b. "Breaking the Link Between Food and Biofuels," *Iowa Ag Review* 14(3). http://www.card.iastate.edu/iowa_ag_review/ summer_ 08/article1.aspx

Babcock, B.A. 2008c. "How Low Will Corn Prices Go?" *Iowa Ag Review* 14(4). http://www.card.iastate.edu/iowa_ag_review/fall_08/article1.aspx

Babcock, B.A., and C.E. Hart. 2008. "Options for the Conservation Reserve Program," *Iowa Ag Review* 14(2): 6-7. http://www.card.iastate.edu/ iowa_ag_ review/spring_08/article3.aspx

Baker, Justin, and Christopher Galik. 2009. "Policy Options for the Conservation Reserve Program in a Low-Carbon Economy." Climate Change Policy Partnership working paper, June. http://nicholas. duke.edu/ccpp/ccpp_ pdfs/low.carbon.policy.pdf

Biomass Research and Development Initiative (BRDi). 2008. *Increasing Feedstock Production for Biofuels: Economic Drivers, Environmental Implications, and the Role of Research*. Biomass Research and Development Board, Dec.

Eve, M.D., M. Sperow, K. Howerton, K. Paustian, and R.F. Follett. 2002. "Predicted impact of management changes on soil carbon storage for each cropland region of the conterminous United States," *Journal of Soil and Water Conservation* 57(4): 68a-73a.

Johansson, R., M. Peters, and R. House. 2007. *Regional Environment and Agriculture Programming Model*. U.S. Department of Agriculture, Economic Research Service, TB-1916, March.

Kirwan, Barrett, Ruben N. Lubowski, and Michael J. Roberts. 2005e. "How Cost-Effective Are Land Retirement Auctions? Estimating the Difference between Payments and Willingness to Accept in the Conservation Reserve Program," *American Journal of Agricultural Economics* 87(5): 1239-1247.

Lubowski, Ruben N., Andrew Plantinga, and Robert Stavins. 2008. "What Drives Land-Use Change in the United States? A National Analysis of Landowner Decisions," *Land Economics* 84(4): 529-550.

Parks, P.J., and I.W. Hardie, eds. 2003. *The Economics of Land Use.* International Library of Environmental Economics and Policy. Aldershot, UK, and Burlington, VT: Ashgate.

Pineiro, Gervasio, G. Esteban, Justin Baker, Brian C. Murray, and Robert B. Jackson. 2009. "Set-Asides Can Be Better Climate Investment Than Corn Ethanol," *Ecological Applications*, 19(2): 277-82.

Reuters. 2008. "Lower enrollment cap likely in US farmland reserve," *Reuters News*, 3 April.

Secchi, Silvia, John Tyncal, Lisa A Schulte, and Heidi Asbjornsen. 2008. "Raising the Stakes," *Journal of Soil and Water Conservation* 63(3): 68a-73a.

Secchi, Silvia, Phillip W. Gassman, Jimmy R. Williams, and Bruce Babcock. 2009. "Corn-Based Ethanol Production and Environmental Quality: A Case of Iowa and the Conservation Reserve Program," *Environmental Management* 44: 732-744.

Soil and Water Conservation Society (SWCS) and Environmental Defense Fund. 2008. *Conservation Reserve Program (CRP): Program Assessment.* April. http://www.swcs.org/documents/filelibrary/CRPassessment report_ 3BE FE868DA166.pdf

Trostle, Ronald. 2008. "Fluctuating Food Commodity Prices: A Complex Issue With No Easy Answers," *Amber Waves*, Nov. http://www.ers.usda.gov/amberwaves/november08/PDF/FoodPrices.pdf

End Notes

[1] A summary of the CRP's enrollment, with an overview of the program's environmental impact, can be found at USDA's Farm Service Agency website: http://www.fsa.usda.gov/FSA/webap p?area=home&subject=copr&topic=r ns-css

[2] Figure 1 charts the timeline of yearly CRP acres, yearly expenditures, and yearly continuous CRP acres. For further detail on the history of the CRP, see http://www.ers.usda.gov/Briefing/ ConservationPolicy/retirement.htm or http://www.fsa.usda.gov/FSA/webap p?area=home&subject=copr&topic=r ns-css.

[3] The EBI is an additive index that includes both environmental components and cost components. High-cost offers receive fewer points in the cost component. Hence, all else held equal, more productive land (that is, likely to have high offer prices) will have lower EBI scores and is less likely to be accepted into the program. Note that cost enters "linearly"—the EBI is not used to create a cost-benefit index. This tends to reduce the positive (negative) impact on EBI scores of very low (high) bids.

[4] In the large 1997 signups, about 60 percent of the 22 million acres offered to the program was from already enrolled land. However, such lands were not given precedence—the EBI scores were used to rank all offers.

[5] USDA's National Agricultural Statistics Service (NASS) reports a rent production index (with a 1990 value of 100) that reached 136 in 1997, and then varied between 113 and 129 for the next several years. In 2006 and 2007 this index rose, reaching 151 in 2007 (http://www. nass.usda.gov/Publications/Ag_Statistics/2008/CHAP09. PDF). Average corn prices were $2.43/ bushel in 1997, ranged from $1.94 to $2.42 between 1998 and 2005, and then rose to $3.04/bushel in 2006 (http:// www.nass.usda.gov/Publications/ Ag_Statistics/2008/ Chap01.pdf).

[6] Abbot et al. (2008) identified three broad forces driving food price increases: change in production and consumption of key commodities, the depreciation of the dollar, and growth in biofuels production. Trostle (2008) identifies a similar set of factors and also notes adverse weather and policy responses.

[7] USDA projections show crop prices staying high through 2020.

[8] For example, USDA's February 2010 Agriculture Baseline Projection predicts fairly constant commodity prices that are greater than prices (between 1998 and 2005) when most current CRP contracts were enrolled (http://usda.mannlib.cornell. edu/MannUsda/view Document Info. do?documentID=1192).

[9] While USDA currently allows the sale of carbon offset credits on land enrolled in the CRP, existing and proposed carbon "cap-and-trade" programs generally consider CRP land ineligible to participate in carbon offset sales.

[10] Note that 2007 prices are similar to prices seen in the spring of 2010.

[11] The choice of 15 billion gallons of crop-based ethanol is predicated on targets set in the 2015 Renewable Fuel Standard. In this research, we abstract from biodiesel, cellulosic, and other varieties of biofuels.

[12] These scenarios focus on the impacts of possible price increases. USDA's February 2010 projections (http://usda. mannlib.cornell.edu/MannUsda/viewStaticPage.do?url=http:// usda.mannlib. cornell.edu/usda/ers/94005/./2010/ index.html) predict a mild, short-term decline in commodity prices. However, these projected prices stay near the price levels used in our MediumPrice scenario.

[13] The LTB identifies about 212 million eligible acres, based on cropping history, soil erodibility, and location.

[14] The results from the Baseline suggest that, to the extent the underlying simulations are biased, so too will be our simulated results. However, since these biases are constant across simulations, comparisons of differences across simulations (where each simulation reflects a different scenario) are likely to reduce these biases. Therefore, the results from the LTB model are best viewed as indicators of broad trends that reveal useful insights.

[15] CRP rental rates are based on a Soil Rental Rate (SRR). The SRR is a parcel-specific maximum bid that the USDA will accept. These values are based on measures of county averages of the rental rate for non-irrigated cropland. In general, the actual rent paid by the CRP is at, or near, the SRR (Kirwan et al., 2005).

[16] The results of other rental rate increases were in line with the 60-percent scenario. For example, with a 120-percent across-the-board increase, results (such as offered acres and average EEBI score) were close to the Baseline but at much higher costs.

[17] When considering a parcel of land eligible for the CRP, high environmental benefits (high EEBI scores) are not necessarily correlated with low soil productivity. However, lands accepted into the program are likely to display a positive correlation between productivity and environmental benefits: since productive lands are likely to have a large offer price, for the EBI score to be sufficiently high the land must also have a high EEBI score.

[18] When per-acre CRP rental rates of the MediumPrice adjusted scenario are used in the HighPrice scenario, 33.6 million acres are offered, and the average EEBI score decreases.

[19] We also investigated a transition from the Baseline to the HighPrice large adjusted scenario. Compared to a MediumPrice adjusted to High-Price large adjusted transition, more acres leave the program (9.8 million acres), and acres that enter the CRP are cheaper but have substantially lower EEBI scores.

[20] While not the most current contract data, March 2008 represents a time of relative stability in CRP contracts. Future analysis that examines re-enrollments and opt-outs of the 39th general signup, held in August 2010, may yield different insights.

[21] The absolute measure (of acres that opt out) is sensitive to a number of assumptions, in particular assumptions about how one models the relationship between the relative profitability of the CRP and the decision to offer land to the CRP. These results use a smoothed relationship that may overpredict the responsiveness of enrollment to profitability of agricultural production.

[22] To reiterate, we abstract from issues of leakage, additionality, and permanence. Also, the CRP's EBI does contain a small carbon sequestration factor. In these models, this factor is kept constant (so that carbon sequestration does not affect EBI scores significantly). The sequestration rate is specific to a multi-State crop production region. It does not depend on what crop is being grown, nor does it incorporate parcel-specific soil information.

In: The Conservation Reserve Program
Editor: Preston L. Cruz
ISBN: 978-1-63482-282-4

Chapter 3

OPTIONS FOR IMPROVING CONSERVATION PROGRAMS: INSIGHTS FROM AUCTION THEORY AND ECONOMIC EXPERIMENTS*

Daniel Hellerstein, Nathaniel Higgins and Michael Roberts

ABSTRACT

The U.S. Department of Agriculture spends over $5 billion per year on conservation programs, mostly on voluntary programs that give financial assistance to farmers and landowners to provide environmental services (such as implementing nutrient management programs or planting native grasses). Since most programs cannot fund all interested parties, program managers must use some mechanism to select applicants. One option is to elicit offers through an auction. This report addresses the use of auctions in conservation programs. It considers how information in the hands of Government officials and rural landowners affects the auction's performance, and how auction design can reduce Government expenditures or encourage landowners to provide greater environmental services. Results of laboratory experiments are discussed, highlighting shortcomings of common features of conservation program auctions (such as limits on the rent landowners may request), as well as how alternative auction designs can improve performance.

---

* This is an edited, reformatted and augmented version of Economic Research Report Number 181, issued by the U.S. Department of Agriculture, Economic Research Service, January 2015.

What Is the Issue?

The U.S. Department of Agriculture spends over $5 billion per year on conservation programs, mostly on voluntary programs that give financial assistance to farmers and landowners to provide environmental services. Most programs cannot fund all interested parties and some use auctions to select from among competing applicants. Auctions are an appealing competitive enrollment mechanism for USDA's conservation programs—they can be cost effective and relatively easy to administer. While well-designed Government auctions can achieve program objectives while utilizing tax dollars efficiently, auctions can also have unintended consequences stemming from the manner in which they are implemented. Auction theory, lessons learned from existing Government auctions, and the results of economic laboratory experiments can all be used to better understand the impacts that auction design can have on outcomes.

What Did the Study Find?

Using the current design of USDA's Conservation Reserve Program (CRP) auction as an example, this report explores how alternative auction designs might provide a bigger "environmental bang for the buck."

Auctions are especially useful when:

1) The buyer (e.g., USDA) can leverage available information about participants to keep costs down;
2) No well-established market exists; and
3) The Government needs a fair and transparent way of selecting participants when budgets are constrained. In reverse auctions, such as the CRP (where there is one buyer and many sellers), competition between participants can improve cost-effectiveness and set a market-clearing price for ecosystem services when private markets do not exist.

Auctions may be less useful when:

1) The buyer has very good information on the sellers' costs and can make efficient purchase decisions by simply making take-it-or-leave-it offers to targeted parcels;
2) Costs and benefits of environmental services vary little across potential participants, so a fixed price could be set and offered to anyone providing the environmental service;
3) The market has too few buyers and sellers for effective competition within an auction environment (e.g., when only a handful of landowners meet program qualifications);
4) (4) Cost-effectiveness is one of several potentially conflicting design criteria; considerations such as income support or broad geographic distribution of participants may be as important.

In these cases, auctions may be no better than administratively simpler approaches, such as offering a single price to anyone wishing to sell or negotiating a seller-specific price. Furthermore, if farmers would have adopted good stewardship practices without financial assistance, any payment mechanism that deters voluntary action can be counterproductive.

When designing a conservation auction, the details matter. Decisions on how to elicit offers, the choice of criteria used for ranking and selecting offers, and the amount of information that will be provided—all of these can affect the auction's performance. Such design elements can also affect who chooses to participate in the auction, how competitive their submitted bids will be, and whether their offers are accepted. All of these outcomes will also influence an auction's cost-effectiveness.

The auction mechanism used in the CRP's general signups, responsible for most of the land enrolled in the program, is a good example. The CRP uses a parcel-specific *Soil Rental Rate* (SRR)—an estimate of the parcel's agricultural rental value—to determine the parcel's maximum acceptable bid, or *bid cap*. While intended to prevent excessive landowner profits, bid caps can have negative consequences. The bid cap is essentially an estimate of the minimum price a seller would accept and still participate in an auction. If the bid cap is less than a seller's true cost of program participation, however, the seller will not make an offer. If dissuaded landowners have eligible parcels with low agricultural value and high environmental benefits, an under-estimated bid cap can lead to higher overall program costs. Bid caps can also discourage landowners from incorporating conservation practices that improve the quality of their offers—even though such improvements are valued highly by the conservation program.

Other types of auctions could do more to limit costs and improve performance. For example, both *quota auctions* and *reference-price auctions* could be implemented with the information currently used by the CRP (the SRR), but neither imposes a bid cap. Quota auctions group similar participants together, and a fraction of the least competitive offers from each group is rejected to induce competition among low-cost participants. This approach requires accepting some higher cost participants into the program, but may be worthwhile if it increases participation and competition among low-cost participants, improving the overall cost-effectiveness of the program. Reference-price auctions assign an estimate of value—a reference price—to each parcel of land. Bids are ranked relative to the reference price—in the CRP example, the reference price could be the SRR. Sellers are penalized if their bids exceed USDA's expectations, but they are not prevented from submitting any bid they wish. Since all bids are considered relative to their SRR, offers with exceptionally high environmental benefits could be accepted even if their bid exceeds their SRR. This could result in improved program cost-effectiveness by both increasing competition among low-cost sellers and increasing the program's overall environmental benefits. While improved performance is not guaranteed, laboratory experiments show that these alternatives can reduce costs up to 18 percent using a reference-price auction, and up to 14 percent using a quota auction.

How Was the Study Conducted?

A literature review highlights the basics of auction theory and design, the advantages and disadvantages of auctions, the different kinds of auction mechanisms that conservation programs could use, and factors to consider when designing an auction. The auction mechanism used by the CRP is examined in some detail, focusing on the impacts of the bid cap. Using actual offers in the CRP's general signup, the limitations and drawbacks of the imposition of bid caps are explored, focusing on how bid caps can dissuade potential participants or the installation of conservation practices that improve the environmental quality of offers. Alternative auction mechanisms are investigated via economic experiments performed in classroom laboratories. These experiments mimic the design of conservation programs to test the potential impacts of alternative auction mechanisms. Regression analyses of experimental results, as well as a numerical simulation model, demonstrate the size of potential gains.

INTRODUCTION

Environmental conservation has been an integral part of U.S. Federal agricultural policy since the Soil Conservation and Allotment Act of 1936. Currently, the U.S. Department of Agriculture (USDA) spends over $5 billion per year on conservation programs.[1] The largest Federal conservation program is the USDA's Conservation Reserve Program (CRP), which pays farmers to voluntarily shift crop acreage from crop production to land covers that conserve soil, enhance wildlife, and reduce erosion.

USDA conservation programs continually change. For example, when the CRP was created in 1985, it focused on reducing soil erosion. Since the mid-1990s, however, the program has considered a broader set of environmental goals, including enhancement of water quality, air quality, and wildlife habitats. Working lands programs—programs that provide financial assistance to farmers to adjust their management practices rather than retire land from production—also continue to expand.[2]

There have also been changes in conservation program implementation, such as how USDA elicits conservation proposals from farmers, decides which proposals to accept, and decides how much to pay participating farmers. These details of implementation, often the result of a mix of legislation and administrative decisions, can be critical to the effectiveness of the program in meeting its goals (Hamilton, 2010). They can help determine which farmers and which lands are ultimately enrolled, which farmers and landowners gain financially from the program and how much they gain, and the level of Government expenditures and environmental benefits associated with the program.

This report addresses how the use of auctions in conservation programs can enhance program efficiency and performance. It considers auction mechanisms used in current and past conservation programs, as well as mechanisms that are more commonly found in other types of auctions. To illustrate tradeoffs involved with different ways of compensating farmers for providing conservation services, the study reports best practices in auction design that apply to the specific case of *reverse auctions*—auctions with many sellers and one buyer—for goods that vary in quality and value (see glossary).

Glossary

Discriminatory Auction: An auction where sellers receive the bid they request. In multi-unit auctions, different sellers will receive different amounts. In contrast, all accepted offers receive the same amount under a uniform price rule (a *uniform-price auction* [see below]).

Economic Rent: The amount of economic profit earned by an activity. In contrast to financial profit, economic profit is net of *opportunity costs* [see below].

Information Rent: The amount of economic rent earned by participating in an auction that is due specifically to the private information that a bidder has. If a bidder earns economic rent by leveraging information about the true cost of participating in a conservation program, then these are information rents. For example, a seller with a true *reservation value* [see below] of $10 who earns a conservation payment of $15 will earn an information rent of $5.

Opportunity Cost: The value of foregone opportunities. If, by participating in a conservation program, a farmer foregoes the opportunity to earn money through the normal operation of the farm, then the money that would have been made through the normal operation of the farm is part of the opportunity cost of participating in the conservation program. Any other foregone benefits (such as the personal benefits of actively farming) are also part of the opportunity cost of participating.

Reservation Value: In auctions, the reservation value of an item refers to the value of the item to the buyer or the seller—the exact maximum amount that a buyer is willing to pay for an item, or the exact minimum amount that a seller is willing to accept for an item. For example, a profit-maximizing seller will enter an auction knowing that he or she will never accept less than his or her reservation value.

Reserve Price: A reserve price refers to a price ceiling or a price floor—a limit on the price of an item to be sold at auction. In an auction with a single seller and many buyers (a forward auction), a reserve price specifies the minimum amount of money that the seller is willing to accept for the item. In an auction with a single buyer and many sellers (a *reverse auction* [see below]), a reserve price specifies the maximum amount of money that the buyer is willing to pay for the item. This is referred to as a bid cap in the Conservation Reserve Program.

Reverse Auction: In a reverse auction, there is one buyer and many sellers.

For example, the Conservation Reserve Program currently uses a reverse auction, with the USDA acting as the buyer and eligible landowners acting as sellers. In contrast, the more familiar forward auctions (e.g., Treasury bond auctions) involve one seller and many buyers.

Uniform–Price Auction: An auction where all sellers receive a single payment which may be different than their actual bid. This payment may be equal to the highest rejected bid, the lowest accepted bid, or some other measure (such as a randomly generated value, with all bids below this accepted). Note that setting a fixed price in a market is not the same as a uniform-price auction; although all sellers receive the same price, setting a fixed price in a market is done without gathering bids in an auction (any eligible seller can chose to participate after the fixed price is announced).

The results of several economic experiments comparing competing auction designs are also presented. These design attributes are chosen in light of current practices at USDA and use the CRP as an example of an auction designed specifically to reduce the cost to the Government of running a conservation program. The experiments reveal the potential effects of alternative auction designs. Taken together, the review of the literature and the laboratory experiments represent preliminary steps in a process that could eventually lead to modifications in current policy.[3] This report does not make policy recommendations on auction mechanisms. Instead, it highlights the feasibility (and limitations) of practical tools that might be used to enroll landowners and farmers in voluntary conservation programs, given what is currently known about auction design.[4]

Auction Basics

Auctions facilitate competition and can lead to efficient price formation.[5] With clear rules regarding participation, the bidding process, and selection criteria, auctions can facilitate simple and transparent enrollment. Many branches of the Government use auctions for various kinds of activities. A few prominent examples (see box, "Examples of Auctions Used Within the Federal Government") include regular sales of Government bonds by the U.S. Treasury,[6] auctions held by the U.S. Federal Communications Commission (FCC) for the telecommunications spectrum, and regular sales of Government property by the General Services Administration. These high-value, high-stakes auctions are remarkable both because of the important roles that they play in allocating public resources and because the coordinating agencies have modified them significantly over the years using a process of experimentation to explore new auction designs.

The design of an auction—how offers are elicited, scored, ranked, and selected, and how prices are set, as well as other details (such as the nature of the information shared by the buyer with sellers)— affects who chooses to participate in the auction, how competitive submitted bids will be, and which offers are accepted. This study focuses on auctions likely to be used by conservation programs— auctions where there are many sellers (such as rural landowners) and one buyer (such as the USDA). This class of auctions is known as reverse auctions.

The design of a USDA conservation auction ultimately determines how well the program achieves its environmental objectives. When designing any auction, information matters—particularly what the buyer knows about the attributes of the sellers. Consider a hypothetical case where the managers of a land retirement program possess perfect information on all eligible landowners. That is, the managers know the exact amount of money it would take to entice each landowner to participate in the program—a figure often referred to as a landowner's *reservation value* or *opportunity cost* (see glossary)—and the exact value society places on the environmental benefits the landowner would provide for that reservation value. Based on this information, a ranking of all eligible parcels could be constructed. Program managers could then choose which parcels to recruit into the program and offer each parcel its reservation value.

Box: Examples of Auctions Used Within the Federal Government

Treasury Auctions

The U.S. Department of the Treasury (Treasury) uses weekly auctions to sell bills and quarterly auctions to sell bonds; the results of these auctions help determine the cost of Government borrowing. The Treasury uses a uniform-price format to sell debt of different maturities—debt instruments with 1 year or less to maturity are known as bills, debt instruments with between 2- and 10-year terms are notes, and long-term debt is issued as bonds.

Many academics and public policy researchers have studied Treasury auctions, and the Treasury has adjusted its procedures as a result of their findings.

Before 1992, Treasury used a multiple-price auction—an auction in which winners paid their bid prices (also known as a discriminatory auction). Due to academic research suggesting that the uniform-price format might lead to higher participation and lower debt-financing costs, Treasury instituted an experiment in 1992 to find out which auction was more cost effective. In the experiment, 2-year and 5-year notes were sold using a uniform-price format, while remaining securities were sold under the old format. Due (in part) to the results of the experiment, the uniform-price auction was implemented for all securities in 1998.

Spectrum Auctions

The Government allocates spectrum for telecommunications in another well-studied auction. In the 1950s, Herzel (1951) and Coase (1959) argued that auctions should be used to allocate the right to broadcast in certain ranges of the spectrum. Until 1981, the U.S. Federal Communications Commission (FCC) assigned spectrum based on cases made to the FCC by applicants; from 1981 to 1993, spectrum was allocated by lottery. In 1993, Congress directed the FCC to auction spectrum and, in 1994, the FCC held its first auction for wireless spectrum, the Nationwide Narrowband Personal Communication Service (Narrowband PCS) auction.

The FCC did not have a history of successful sales to draw on in order to decide what auction format worked best. Laboratory experiments were conducted at the California Institute of Technology, and a smaller July 1994 auction (worth $617 million) informed a subsequent December auction (worth $17 billion).

In reality, information on eligible parcels is far from perfect. While measures that are predictive of reservation value (such as soil quality and local agricultural rents) may be available, these are not perfect predictors of a given parcel's value to its owner. Furthermore, collection of detailed information is costly. Careful auction design can leverage available information to achieve cost-effective results.

The amount of information available influences landowners as well as the Government. If landowners know the goals and budget of a conservation program and are aware of the reservation values of other parcels, they could strategically adjust their bids (Cason and Gangadharan, 2004). These bid adjustments could take the form of individual actions (such as owners of low-valued parcels demanding a high price) or even collusive behavior that could undermine the goals of the auction. However, auction design choices can also mitigate these risks.

LESSONS FROM AUCTION THEORY AND PRACTICE

Economists and scholars from other disciplines have studied auctions by examining bidding patterns observed in actual auctions and the results of auction experiments,[7] which provide some broad lessons that may be useful in the design of conservation program enrollment schemes. There is also a growing body of literature that studies the practical design of markets, especially markets created by Government entities, which offers lessons that are applicable to the design of Government auction markets for environmental services.

The Advantages and Disadvantages of Auctions

Auctions can offer several advantages over negotiations or fixed-price offers:[8]

1) Auctions reduce costs to the buyer by using competition to drive down prices. This is especially true when the buyer can leverage information about participants to limit *information rents*. Information rents are profits a seller obtains due to proprietary knowledge (say, about the seller's cost of production) that the buyer cannot observe (see glossary).
2) Auctions are effective at setting a price when the Government is uncertain of what price to pay. This frequently occurs when the Government is purchasing goods that have a limited number of buyers and sellers. Uncertainty is resolved through the price revelation (e.g., the price at which a buyer is willing to buy and a seller is willing to sell) that happens naturally in auctions between sellers and the Government. In comparison, bilateral negotiations (e.g., bargaining with individual landowners) can be a time-consuming and expensive means of information discovery. Auctions can also be an effective means of discovering a market-clearing fixed price, which may be used in future acquisitions (Athey et al., 2002).

3) Auctions are a fair and transparent way to allocate goods when rationing is necessary. When budgets are constrained (e.g., when a program is oversubscribed) the Government must choose whom to buy from and at what price. An auction resolves this problem by specifying a general set of rules before the auction occurs and evaluating potential offers according to these rules.

However, auctions are not always a superior mechanism. There are several factors to consider when deciding if an auction is the right procurement mechanism for a particular program:

1) Homogeneity in costs or environmental quality may reduce the advantage of auctions over alternative procurement systems. When the pool of suppliers is homogeneous, a properly chosen fixed-price mechanism will have a similar outcome to an efficient auction.
2) Auctions become more administratively expensive when parcels have unique benefits. If candidate parcels are wholly unique (such that the parcels have no close substitutes), a more explicit targeting mechanism that identifies desired parcels might be more suitable. Auctions can be used for procurement in these situations, but selecting offers becomes a more complex process that includes a mechanism to weigh quality.[9] This process requires potentially costly input from the Government (see box, "Scoring Auctions").
3) If adopting a payment mechanism substantially deters prior voluntary actions, any payment mechanism—whether based on a market, an auction, or a Government program—may be counterproductive.[10] Similarly, auctions may not be cost effective if they end up enrolling those who would have undertaken a desired course of action without any payment (Arnold et al., 2013).

Box: Scoring Auctions

In Government procurement, reverse auctions can be used to procure many goods, including both commodities (goods of uniform quality) and highly differentiated goods (such as construction contracts or contracts to provide environmental services). Scoring auctions take quality into account when selecting the winning offers. In a scoring auction, a numerical score is assigned to each offer, offers are sorted by the score, and the winning offers are those with the best scores.

The Conservation Reserve Program is an example of a scoring auction—offers are sorted according to the Environmental Benefits Index, a composite score that takes both the cost and the environmental implications of each offer into account.

The formation of the scoring mechanism is an extremely important part of auction design in a market for goods of heterogeneous quality. The score is a numerical representation of the buyer's preferences, including the relative preference of the buyer for various measures of quality and price. A scoring mechanism that is poorly designed can lead to the selection of offers that fail to achieve program objectives and the rejection of offers that would have yielded preferable results.

Construction contracts are another prominent example of quality-differentiated goods that are awarded by scoring auctions. Lewis and Bajari (2011) studied construction contracts awarded by the California Department of Transportation (CDOT), some of which used a scoring function of $b + c*d$, where both price (b) and days-until-completion (d) are considered. A value c equal to 10,000 means that contractors can lower their score—make their score more competitive— equally well by reducing their requested payment by $10,000 or by shortening their contract term by 1 day. By increasing c, CDOT implicitly increases the public's preferences for shorter contracts by equating a reduction in the term of the contract with a larger reduction in price. For more information on the use of scoring auctions, see the recent report by the U.S. Government Accountability Office (2013) and Asker and Cantillon (2008).

4) When accurate price information is available, auctions are less informative than they would be in a low-information environment. If the purchaser can observe characteristics (including costs) with a high degree of accuracy, auctions may not be more cost effective than more direct mechanisms (such as the Government making direct offers to sellers).
5) If public demand for the procured goods is large relative to supply, competition in an auction will be low and pressure on participants to place bids close to their opportunity costs will be reduced, thereby limiting the ability of auctions to reduce information rents. The expense of negotiations may be justified if competition cannot be relied upon to control costs.

6) The goals of the program must be compatible with competition. If cost-effectiveness is not a primary goal (e.g., if income support is the primary goal of the program) auctions may be inappropriate.
7) Efficient allocation is not always achieved by an auction. In general, social welfare—the value of goods and services produced by a society without concern for their distribution—can be maximized by setting a price equal to the social value of the good being procured and accepting all interested offers at that price. For example, if the managers of a conservation program know the monetized value of the reduction in soil erosion due to retiring an acre of farmland, a uniform price equal to this amount could be announced and all offered acres accepted. Since landowners willing to sell at this price will have an opportunity cost less than this price, total social welfare will be maximized.[11] In contrast, cost-effective auctions might entail rejecting some low-cost sellers to motivate competitive bidding by all participants.

   However, if budgets are constrained and the Government is not capable of purchasing from all willing sellers at the socially efficient price, reducing procurement costs will free up money to purchase more units. As detailed in the box, "Economic Efficiency Under Limited Budgets," the social welfare gain from purchasing extra units can then justify the use of a least-cost procurement mechanism, even in a world where efficient allocation is important.

Designing an Auction

There are several considerations when designing an auction, such as defining the objective of the auction, defining the good being auctioned and communicating the rules of the auction (see box, "Auction Design—Primary Steps").[12]

Other basic decisions include *uniform-price* versus *discriminatory*[13] auctions (see glossary), sealed versus observable bids, and the use of bid caps (maximum allowable bids) (see box, "Kinds of Auctions"). While these steps apply to all auctions, the characteristics of auctions most likely to be relevant for the procurement of environmental goods and services have further distinctions.

Box: Economic Efficiency under Limited Budgets

An economically efficient level of provision of a good occurs when its marginal benefit is equal to its marginal cost—where the value of the last unit consumed equals the expense of supplying it. Applying this idea to agricultural conservation, the conservation benefits from the last acre of land accepted into a conservation program should equal the best alternative use of the land. This outcome can be achieved using a simple market—where a single price is announced and all sellers willing to accept this price are accepted.[1] However, the optimality of a simple market depends on several assumptions. In particular, the purchaser (for example, the Government) should have no budget constraint.

This is not often the case, however. For example, an agricultural conservation program may have a fixed budget that is not large enough to purchase the socially optimal quantity of retired cropland. In this case, an auction mechanism—by allowing the available budget to purchase a quantity closer to the social optimum—can yield social welfare outcomes that are superior to a simple market. The following land retirement example illustrates this case. Assume that:

- There are 20 1-acre parcels that have opportunity costs (net profits from agricultural production) between \$2 and \$40.
- The marginal value (benefits of ecosystem services) of retiring parcels is decreasing: the first acre retired provides \$62 in environmental benefits, while the 20th acre retired provides \$24.
- The marginal benefit and cost curves are linear.

The table below shows costs and benefits. The socially optimal price is \$32, with a total expenditure of \$512 and a total benefit of \$900 (the sum of conservation benefits from retired land plus the sum of net agricultural profits of nonretired land).

Suppose, however, that the Government is constrained to spending only \$338. If a simple market mechanism is used, only 13 acres can be retired at a uniform price of \$26. The imposition of this budget constraint drops the net social benefits from \$900 to \$888. In contrast, consider a discriminative auction (that pays selected offers their bid) with a bid cap always \$9.15 greater than the opportunity cost, and assume that all farmers make an offer equal to their bid cap. \$338 will allow 14 acres to be retired, yielding total benefits of \$896.

Alternatively, retaining net benefits at $896 (by retiring 13 acres) would cost $301 under the discriminative auction (89 percent of the $338 in expenditures under a simple market).

The discriminative pricing obtained through the auction mechanism reduces payments to low-cost sellers. These savings are then used to retire more acres, allowing the Government to purchase a number of acres closer to the optimum quantity. While this example's budget constraint (of $338) and bid cap (that is always $9.15 greater than the opportunity cost) were chosen for numerical simplicity, it does illustrate how auctions can yield better social welfare outcomes than simple market mechanisms.

Table. Results from three budget and pricing scenarios

Procurement mechanism	Acres retired	Opportunity cost of retired land	Conservation benefits from retired land	Payments to retired land	Total benefits
Using a socially optimum, uniform price of $32	16	$272	$752	$512	$900
Using a uniform price of $26, given a budget constraint of $338	13	$182	$650	$338	$888
Using a discriminative auction and a budget constraint of $338 (max payment is $37)	14	$210	$686	$338	$896

Note: Total benefits is the sum of conservation benefits from retired acres and the value of production (the opportunity cost) from nonretired acres.

Source: USDA, Economic Research Service.

[1] In more complex situations, such as when conservation benefits differ across parcels, setting a single price may not be economically efficient. In these situations, other enrollment mechanisms can be used to maximize social welfare. For example, Polasky et al. (2014) introduced a subsidy auction, which achieves social efficiency by separating bids from payments (which incentivizes submission of bids equal to actual opportunity costs). However, the subsidy auction achieves efficiency by awarding all the net benefits of conservation to the parcel owners, and it assumes an unlimited budget.

Box: Auction Design —Primary Steps

Although the details of an auction matter, a great deal of practical experience suggests that good auction design begins even before detailed decisions are made. The following primary steps are also important to successful auction design (Depiper et al., 2013):

- Define the objective of the auction from the perspective of the buyer.
- Clearly define the good being auctioned. Design the auction to be appealing to potential sellers (more sellers translates to more competition) and safe for sellers to participate in (participating should not put sellers at undue risk).
- Clearly communicate the rules of the auction to all potential sellers. Sellers should feel confident that they know how they can improve their offers, and they should understand basic tradeoffs.

Box: Kinds of Auctions

There are four main types of auctions used to buy or sell single items.[1] Since there is no important conceptual difference between reverse auctions (one buyer, many sellers) and forward auctions (one seller, many buyers), the following descriptions focus on reverse auctions.[2]

First-price (or "low-price") sealed-bid auctions	Each potential seller offers a private (unannounced) bid to the buyer. The buyer selects the seller with the lowest bid and pays the amount of that bid for the good.
Second-price sealed-bid auctions	Each potential seller offers a private bid to the buyer, and the buyer selects the seller with the lowest bid. Unlike first-price sealed-bid auctions, however, the winning bid does not determine price. The buyer pays the winning seller an amount equal to the second-lowest bid.
Descending auctions	A starting price is announced. This price is sufficiently high that all sellers are willing to sell. Price is lowered successively until only one seller remains. The winning seller is paid the final price (the price at which he or she was the only willing seller).
Ascending auctions	The buyer begins by announcing a very low price and gradually raising it. The first seller to announce that he or she will accept the announced price supplies the good at that price.

When multiple units are bought (or sold) in a single auction, analogs of the above auction types exist:

Discriminatory sealed- bid auctions, or pay-as- bid auctions	Each potential seller offers a private (unannounced) bid to the buyer. The buyer orders bids from lowest to highest and purchases the lowest bids until he or she has purchased the desired amount. The buyer pays each winning seller an amount equal to his or her bid.
Uniform-price sealed- bid auctions	Each potential seller offers a private bid to the buyer and the buyer selects the lowest bids. However, the winning sellers are not paid according to their own bids. The buyer pays each of the winning sellers the same price. The price is usually determined by one of two rules: the highest bid submitted by a winning seller or the lowest rejected bid.
Descending auctions	A starting price is announced. This price is sufficiently high that, in total, sellers are willing to supply more good than the buyer is willing to purchase. Price is lowered successively until, as sellers drop out, the amount of supply offered by willing sellers is equal to the demand of the buyer. The winning sellers are all paid the final price (the price at which supply equals demand).
Ascending auctions	The buyer begins by announcing a very low price and gradually raising it. Sellers join as the price increases until the amount of supply offered by willing sellers is equal to the demand of the buyer. The winning sellers are all paid the final price (the price at which supply equals demand).

[1]See Menezes and Monteiro (2005) and Krishna (2010) for extensive discussion of basic auction varieties.

[2]The forward-auction analogs of these reverse-auction types are known as a first-price auction, a second-price auction, an English auction, and a Dutch auction, respectively.

Observable Differences in Costs among Sellers

One benchmark assumption in auction theory is that all potential sellers have private opportunity costs drawn from the same distribution. This means that, while some potential sellers may have higher costs of providing the good than others, neither the buyer nor the other sellers can observe these differences before the auction takes place. Each seller appears identical (symmetric) from the perspective of the buyer and from the perspective of all other sellers. In contrast, auctions in which there are clear observable differences between sellers—auctions where meaningful opportunity-cost differences can be predicted using observable information—are called asymmetric.

In conservation and environmental services auctions, an assumption of symmetry is often false— instead, the assumption of asymmetry is an important determinant of good auction design. The costs of providing environmental services can vary widely depending on where land is located and the agricultural or other economic activity that would otherwise take place on the land. While some of the cost differences between sellers may remain unobservable, many of these differences are clearly evident to all sellers and to the buyer. For example, it is commonly known that the opportunity cost (the lost revenue per acre from not producing crops) to a farmer of establishing native grasses on an acre of wheat in Montana is less than on an acre of corn in Iowa.

Observable differences among sellers may deter sellers from participating in an auction. Low-cost sellers (such as the Montana farmer with a low opportunity cost) are generally willing to accept lower prices than high-cost sellers (such as the Iowa farmer with a high opportunity cost); low-cost sellers expect to be accepted into the program while high-cost sellers expect not to be accepted. If sellers have a good understanding of others' costs, then high-cost sellers have little incentive to bid in the first place, especially if the act of bidding is costly. Low-cost sellers may recognize that high-cost sellers will not submit bids, and increase their bids accordingly.

More broadly, the fact that sellers are observably different should be considered in the design of an auction. Treating all potential sellers as identical can appear even handed but may deliver high profits to sellers with observably lower opportunity costs. Using observable information may help the Government construct an auction that gives all auction participants a chance to make some profit, but which reduces the large profits that low-cost sellers would otherwise enjoy.

Recurring Auctions and Repeat Participation

The vast majority of auctions have been modeled and analyzed as single auctions: the auction happens, a transaction occurs, and the process ends. There is no interaction between the market and the auction, and no opportunity to participate in the auction again. Many conservation and environmental services auctions, however, are conducted on a regular basis (Latacz-Lohmann and Schilizzi, 2007), which creates opportunities for two types of important seller behavior—choosing when to bid and learning how to bid well.

If potential sellers regard the opportunity to bid as a recurring opportunity, they may delay bidding until they feel that the process will be most valuable to them. The value of waiting could be caused by strictly external factors, such as the opportunity to use land for an alternative purpose (i.e., during a period of high crop prices), or external changes in the structure of the program that make participation more or less favorable (i.e., congressional action to increase or decrease funding to the program). The value of waiting to bid is also influenced by the ability to learn from the outcomes of earlier auctions. By observing the outcomes of an auction today, a seller can learn how to modify his or her behavior to maximize profits in an auction tomorrow.

The ability of sellers to participate in the auction of their choice, and learning how to bid well in that auction, eliminates some of the monopsony power of the buyer. Informed sellers will use all the information available to them to extract profits from the auction, which may serve to erode several of the traditional cost advantages of auctions (Latacz-Lohmann and Schilizzi, 2007).

Entry

Encouraging sellers to enter an auction is always a key consideration of auction design. Without broad participation, an auction will not be competitive and, at the very least, the lack of competition erodes one of the major motivations for using an auction mechanism in the first place. In extreme cases, a lack of participation can lead to under-enrollment and an auction that does not meet its primary policy goals (e.g., retiring a targeted amount of highly erodible cropland).

Because conservation auctions frequently rely on the participation of a large number of individuals, many of whom are not necessarily experienced sellers, it is especially important that information about the auction is widely available to all potential participants. In fact, promoting the opportunity to participate is a key factor in good conservation auction design. In some cases, disseminating detailed information about the auction could enhance competition by lowering barriers to entry for inexperienced sellers.

An extreme case of this type of information campaign can be seen in the FCC's spectrum auctions, as well as the planned auction of takeoff and landing slots at major airports by the U.S. Federal Aviation Administration (FAA). In both instances, the Federal agency in charge of the auctions gave potential sellers the opportunity to participate in mock auctions so that they could learn about the auction system.[14] Potential sellers gained experience with the relevant software, asked logistical questions of the buyer, and could test bidding strategies in a setting without consequences. Additionally, the auction for agricultural irrigation rights held by the State of Georgia also used practice auctions to spread information and allow potential sellers to gain experience—in this case, farmers who were likely inexperienced with the particular mechanism chosen by Georgia for its natural resource procurement (Cummings and Holt, 2004).

Ambiguity

If auction rules are unclear, potential sellers could have difficulty assessing their best offers. The more complex and ambiguous the rules, the more difficult it will be for sellers to decide how to structure their offer. Substantial literature in behavioral economics indicates that individuals facing such complex decisions will tend to forgo making them and stay with a status quo alternative. For example, many employees choose not to enroll in subsidized retirement savings plans, choosing a default option of saving nothing instead (Congdon et al., 2011). More recent literature documents how simplifying decision environments can aid decisionmaking (see Beshears et al. (2013) for an example where simplifying retirement savings decisions can drastically change choice).

In the case of reverse auctions for environmental services, too much complexity may discourage participation or limit the effectiveness of participants. For example, in a land-retirement program auction, if a seller does not add an extra conservation practice to a field (because he or she does not understand that this is the least costly way to improve the offer), both the seller and the buyer are affected by the result. Ultimately, while some degree of complication is inevitable in many auctions, it is important to reduce the decisionmaking problem facing potential sellers so that it is manageable (i.e., so that the auction format clearly incentivizes behavior that the Government wants to encourage).

Risk and Regret

Sellers who feel they may expose themselves to risk by participating in an auction may be less likely to participate, so an auction design that reduces unnecessary risks to potential sellers will encourage participation. Though it is impossible to eliminate all the associated risks, auction outcomes can be improved when risks can be decreased without harming the goals of the auction. For example, a landowner offering to participate in a multiyear program to improve bird habitat might give up the opportunity to hay grassland (leaving grassland largely untouched is beneficial to the provision of environmental services). At the same time, however, participation in the program exposes the landowner to risks, such as drought that affects the farm's other haying land. Rules that permit infrequent haying or haying in exigent circumstances without harming the provision of environmental services can mitigate the risks of participating in the program.

Common Values

The nature of the costs of providing environmental services, particularly how costs are related among potential sellers, should be a consideration in conservation auction design. A seller's cost may be private or common. When the determinants of the cost of providing environmental services for one landowner are completely unrelated to the determinants of cost for other landowners, costs are said to be private; when the determinants are shared, costs are said to be common. In reality, costs frequently have characteristics that are a mixture of private and common. For example, the cost of providing bird habitat by planting native grasses (an environmental service) rather than producing corn on a parcel of land are determined by both private and common factors. Each landowner faces the same market price of corn (a common determinant of the opportunity cost of not producing corn), but the amount of corn that can be produced depends on soil characteristics that are unique to a landowner's parcel (a private determinant of opportunity cost).

When a large proportion of the determinants of cost are common, winning sellers may be susceptible to the "winner's curse" (Thaler, 1988). That is, when costs are entirely common, winning sellers might actually *regret* having won the auction.[15] This happens if winning sellers (those who bid the lowest) undervalue the opportunities they are foregoing by agreeing to participate in the conservation program. Auction design can mitigate this effect by ensuring that all potential sellers have easy access to public information (such as price forecasts and knowledge of alternative program opportunities)—this can create expectations that are more closely aligned with auction outcomes, which can encourage participation and increase competition and auction effectiveness (Ozbay, 2007).

Collusion and Threats

Important practical considerations when designing auctions include preventing collusion among sellers (e.g., a secret agreement among sellers to limit competition) and deterring outside punishments by rival sellers for competitive bidding within the auction. These concerns typically apply to auctions with few sellers. While it is important for auction design to reduce opportunities for collusion and threats, these are small concerns for large-scale conservation programs because there are typically a large number of geographically dispersed landowners interested in participating.

CASE STUDY: AUCTIONS IN THE CONSERVATION RESERVE PROGRAM

Enrollment mechanisms for many conservation programs already work in a manner similar to reverse auctions: bids or proposals are elicited from many farmers and landowners, and the Government scores, ranks, and selects which proposals to accept. The enrollment mechanism used by CRP, the Nation's largest conservation program, is a prime example of how reverse auctions are used in conservation programs.[16]

Established in 1986, the CRP retires highly erodible and other environmentally sensitive cropland from production.[17] With the program using 10- to 15-year contracts, enrollment peaked in 2007 at 36.8 million acres. As of September 2014, about 25.6 million acres were enrolled at a cost of about $2.0 billion per year (fig. 1). Most of these acres were enrolled in 262,000 competitive generalsignup contracts. However, about 20 percent (5.7 million acres) were enrolled in 410,000 continuous-signup contracts that target environmentally sensitive land (such as stream buffers) through a noncompetitive, "first-come, first-served" selection process.

Offers for enrollment in the CRP consist of a request by the landowner for an annual rental payment in exchange for installing a proposed cover practice (such as installing native grasses or longleaf pine) on a specific parcel of land. CRP participants must pay substantial penalties if they withdraw before the expiration of their contracts (which have a 10- to 15-year duration). Rental requests may not exceed a parcel-specific bid cap, which is set by the Government based on the geographic location of the parcel and the productivity of its underlying soils.[18]

Since 1996, CRP offers (at the national level for the general signup) have been ranked according to the Environmental Benefits Index (EBI),[19] which accounts for the environmental services expected to be produced by the retired parcel of land and the cost of the landowner's requested rent. Offers with the highest EBI scores are accepted into the program.

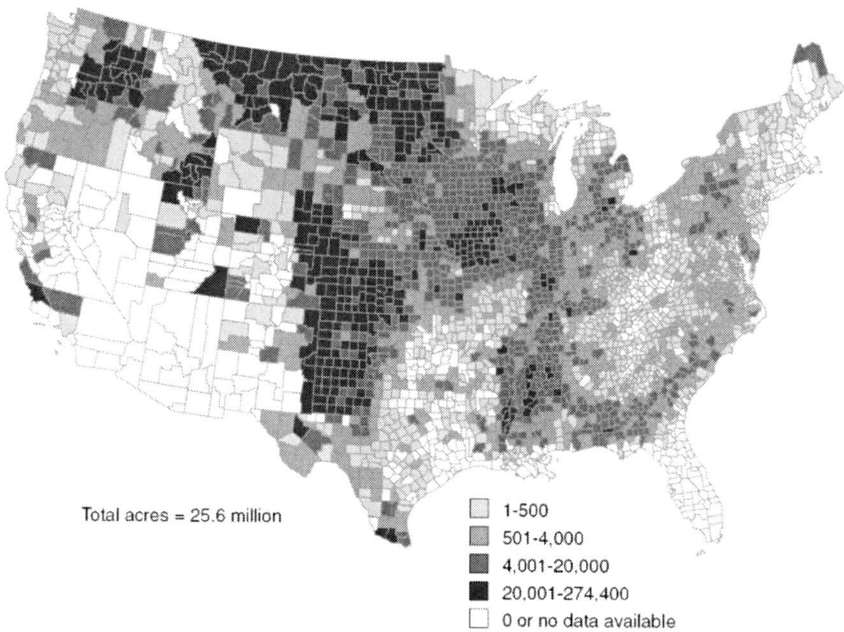

Source: USDA, Economic Research Service using confidential program data from USDA, Farm Service Agency, Conservation Reserve Program.

Figure 1. Conservation Reserve Program enrollment, acreage per county, September 2014.

As with most conservation program auctions, landowners interested in participating in the CRP have a delicate decision to make when deciding how much compensation to request and what cover practice to install—the greater the compensation requested (and the less valuable the proposed cover practice), the lower the odds their offer will be accepted. Thus, a landowner wishing to maximize gains from offering land to the program will balance the desire for a higher payment (and less upfront expenditures on installing cover practices) against the lower odds of having the offer accepted.

The design elements of the CRP auction greatly influence which parcels are selected for the program, the amount landowners are paid, and, consequently, the success of the CRP in achieving its conservation objectives. Consider, for example, four key features of the CRP auction:

1) Contract length. The 10- to 15-year contract required in CRP has costs and benefits. A long-term contract can be beneficial from an environmental standpoint if it takes time to establish native covers,

wildlife habitat, or buffer strips to accrue environmental benefits. On these grounds, contracts longer than 10-15 years may be justifiable.[20] On the other hand, landowners may require greater rental payments if they are locked into the program for a longer period of time (i.e., if they value flexibility).

2) The Environmental Benefits Index. The EBI is a formula that scores each parcel offered for enrollment into CRP. This scoring mechanism, perhaps the most crucial design element of the CRP, weighs environmental and cost-related priorities. Scoring implicitly values each of the program's environmental goals individually.
3) One-time sealed bids. The CRP uses general signups to enroll new acreage.[21] Each general signup uses a discriminatory sealed-bid auction where landowners have a single opportunity to privately submit their bids (bids are not announced publicly) which cannot be revised. This feature of the enrollment process also involves tradeoffs. If landowners could revise their offers after observing a first round of bids, those not accepted in an initial round may be willing to adjust their cover practices or rent requests. However, if landowners knew additional rounds of bidding would occur, they would have less incentive to bid competitively in the initial round.
4) Bid caps. Bid caps are based on an estimate of a parcel's agricultural rental value (its Soil Rental Rate (SRR)), and can have several powerful effects on the CRP. The CRP's bid cap is meant to prevent excessive rental payments. However, bid caps are not perfect measures of opportunity cost, and their use can have several undesirable side effects:
   i. Landowners with parcels that are likely to be accepted (such as highly erodible land) are likely to place a bid equal to the bid cap—they have little incentive to ask for less.
   ii Furthermore, landowners whose offers are likely to be accepted have little incentive to install EBI-improving practices (such as wildlife-friendly cover).
   iii. Landowners, regardless of their EBI scores, are highly unlikely to offer their land if the bid cap is less than their opportunity costs (i.e., what they could earn from their usual farming activities). As a result, even if their land would be more valuable (say, on a benefits-per-dollar basis) to the program than other land that is accepted into the program, the bid cap means these parcels will not be available to program managers for evaluation or selection.

For more information on these potential problems, see the box, "Drawbacks of the Bid Cap." Possible solutions to these bid cap issues and their effects on program participation, cost, and environmental quality are in the "Experiments on Auction Performance Using Alternative Designs" chapter of this report.

Box: Drawbacks of the Bid Cap

Bid caps can prevent excessive overpayment, but they have several drawbacks. We consider three side effects, as mentioned in the report.

Point (i) notes that landowners whose probability of acceptance is high will ask for their bid cap, even if it is well above their opportunity costs. Thus, for parcels that are likely to be accepted— perhaps because they are highly erodible and have low soil productivity—the cap may limit excess profit, but may not eliminate it.[1] This tendency can be illustrated by examining the *automatic* EBI points of actual CRP contracts.[2] These are points a parcel will receive regardless of what the landowner chooses to ask for—they are points that are not affected by the landowner's asking price or the conservation practice the landowner commits to installing. Since parcels with a high number of automatic points are more likely to be accepted, they can be offered with a higher bid than other parcels (all else being equal).

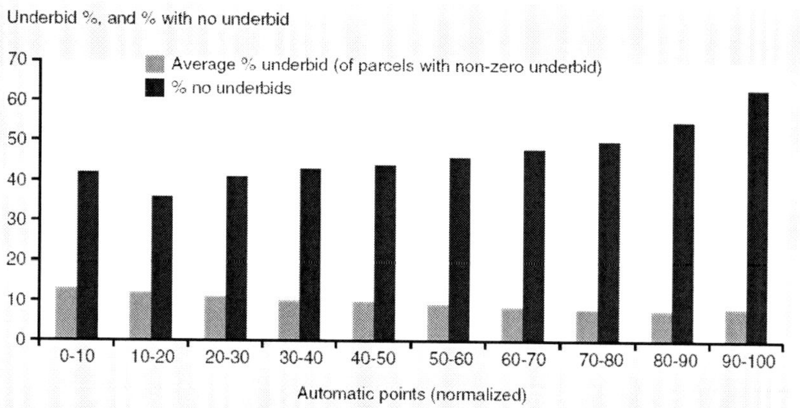

Notes: Each bar represents 10 percent of the offers, where offers are grouped by the normalized automatic points. Automatic points range between 0.1 and 0.89; with a 5-percent to 95-percent range of 0.25 to 0.57. The average percent underbid is for parcels with nonzero underbids.

Source: USDA, Economic Research Service using confidential program data from USDA, Farm Service Agency, Conservation Reserve Program, 1995-2013.

Figure. Underbidding and automatic points.

An examination of CRP contracts confirms this tendency. A 1-percent increase in automatic points will, on average, reduce a parcel's underbid (the difference between the bid and the bid cap) by about 0.12 percent (appendix 2). The figure illustrates that, as automatic points increase, the fraction of participants underbidding decreases, as does the average size of underbids.

Point (ii) notes that landowners with a high probability of acceptance have little motivation to improve their EBI scores by installing higher scoring (but more expensive) conservation practices.[3] If the extra environmental benefits provided by better practices are worth the installation costs, both the landowner and the Government would be better off if the practice was installed.[4] The automatic points of actual CRP contracts can be used to investigate how offer quality is affected by the probability of acceptance. A 1-percent increase in automatic points will, on average, lead to a roughly 0.3-percent reduction in EBI points due to less environmentally beneficial conservation practices being offered (appendix 2).

Last, and most importantly, point (iii) notes how bid caps can negatively affect participation rates. If seller-specific bid caps are set too low, many suppliers of conservation services may not offer their land for enrollment at all. For example, if the bid cap is set equal to the median opportunity cost of potential bidders, half of potential participants will not place offers, even though their parcels may have high environmental values (as reflected in their EBI scores).[5] This will reduce competition among remaining sellers and force the buyer to accept potentially higher bids to meet his goals. As a result, it is quite possible that accurate, but not perfect, bid caps will increase expenditures. For example, Hellerstein and Higgins (2010) present a numerical example illustrating how very accurate, but not perfect, bid caps can have perverse results. They compare expenditures required to achieve an acreage target to expenditures when a less accurate bid cap, that is always noticeably greater than opportunity costs, is used. They find that costs are 48 percent higher in the first (very accurate bid cap) case.

[1] Prior research (Kirwan et al., 2005) estimated that between 10 percent and 40 percent of Conservation Reserve Program expenditures were due to information rent (bids higher than the landowner's reservation price).

[2] Automatic Environmental Benefits Index (EBI) points are awarded based on a parcel's characteristics, including the erodibility of the land, whether it is located in a wildlife priority or a water quality area, and it's Soil Rental Rate (SRR). All other things being equal, parcels with low SRRs will have more automatic EBI points—since they cannot ask for a high bid, the EBI cost factor (that awards more points the lower the bid) will always be large.

[3] For example, rather than using an inexpensive grass cover to prevent erosion, a more costly cover could prevent erosion and also provide high-quality wildlife habitat. Given the USDA's traditionally non-adversarial relationship with farmers, enrollment mechanisms that induce landowners to improve the environmental quality of land enrolled in conservation programs, even without an overall reduction in Government expenditures, may be of great interest.

[4] The money that would be spent on better practices does not disappear—it could go to enroll additional acreages. However, a parcel with a superior practice may be worth more than the additional parcel(s) that could have been obtained. Thus, the drawback of bid caps discouraging installation of better practices matters the most when the environmental value of foregone practices exceeds the marginal environmental benefit of obtaining additional parcels (on a per dollar basis).

[5] If landowners are risk-neutral profit maximizers, and each parcel's Soil Rental Rate is an accurate measure of the median opportunity cost of observationally equivalent parcels, then half of the parcels will have a bid cap that is below their actual opportunity cost. Since landowners of these "higher than bid cap" valued parcels would be financially worse off if their parcels were enrolled, they will not offer them to the Conservation Reserve Program.

AUCTION DESIGN ALTERNATIVES

If cost differences are unobservable to the buyer, the buyer cannot effectively differentiate between sellers; consequently, sellers with very low costs gain much more than other sellers do, especially in a repeated auction where low-cost sellers can better estimate the highest price they can ask for and still have their offer accepted. However, if the buyer has some indication of which sellers have low costs and which have high costs, the buyer can adjust the auction scheme so that low-cost sellers bid less than they otherwise would.

One way to pay different prices to different sellers is to impose seller-specific bid caps that vary according to observed indicators of the seller's cost—such as the SRR used by the CRP. However, as discussed above, tight bid caps have a number of drawbacks. Relieving some of these drawbacks (say, by using an across-the-board increase in bid caps to increase participation) can increase other problems (such an increase in bid caps could end up paying landowners more to do the same thing).

There are other strategies to obtain lower bids from low-cost sellers. A ranking system could narrow the advantage of low-cost sellers when offers are scored and ranked. Low-cost sellers would then face stiffer competition from higher cost sellers who would otherwise be uncompetitive. For example, sellers from areas with higher rental rates might be assigned additional EBI points, or their asking prices could be discounted.[22]

Alternatively, a reference-price auction could assign a nonbinding estimate of value (a reference price) to each parcel of land, with bids ranked relative to the reference price.[23] For example, holding environmental quality constant, a bid of $20 submitted to enroll land with an estimated value of $20/acre would receive a score of 1, while a bid of $20 submitted to enroll land with an estimated value of $30/acre would receive a higher (more competitive) score of 1.5 (i.e., 30/20). Landowners placing bids from lower cost areas (e.g., areas with lower reference prices) would then have an incentive to boost their scores by lowering their bids.[24]

Both ranking systems and reference prices increase the probability that some low-cost sellers will not be accepted and that a high-cost seller might win instead. This feature is meant to induce low-cost sellers to make offers closer to their opportunity cost. In other words, these mechanisms threaten to accept high-cost sellers—sellers that an open auction would not be likely to accept. When successful, such a threat will cause low-cost bidders to submit lower bids, compensating for the increased acceptance of higher cost bidders and thus reducing total costs.

Quota auctions offer another strategy by limiting the number of offers accepted from any group having similar observed characteristics.[25] In an agricultural conservation context, groups could consist of contiguous counties with similar agricultural productivity—in other words, all eligible land within a group will have similar opportunity costs.[26] For example, in a program like the CRP, administrators might limit acceptances to 90 or 95 percent of all offers from any given county (or a USDA crop-reporting district), or place a limit on the absolute number of contracts or acres that can come from this group. Without the limit, landowners in a low-cost group (e.g., a crop-reporting district where agricultural productivity is less than the national average) can submit offers that far exceed their opportunity cost and still be confident of acceptance. A limit on the number, or share, of offers accepted would force landowners to compete with other landowners in their low-cost group, rather than with those (in other groups) with higher costs.[27] This kind of auction can only be used in a multiple-unit setting when there are clear observable differences between sellers, and an ability to readily assign sellers to a group (as is typically the case when buying conservation services across a broad landscape).

Depending on the particular situation, these alternative auction designs might significantly reduce the buyer's costs of procurement.[28] Alternatively, auction design can affect conservation policy goals beyond cost minimization. An auction mechanism that increases costs may be more costly on a per acre basis, but it may lead to a greater provision of environmental services—i.e., a sole focus on cost savings may fail to take advantage of opportunities to substantially increase ecosystem services (from better practices on the same total acreage). For example, even a well-designed bid cap for CRP offers may fail to incentivize low-cost sellers to improve their offers (say, by installing a wildlife-friendly cover practice).

However, any auction mechanism entails tradeoffs. For example, although the CRP's use of parcel-specific bid caps can dissuade participation by low-cost sellers who happen to have bid caps below their opportunity costs, bid caps do limit the amount of extra profit any one seller can obtain. Any plausible quota auction or reference-price auction necessitates accepting some high-cost sellers, which may more than offset savings from lower bids by low-cost sellers. Moreover, these auctions can also influence the distribution of benefits across sellers. In comparison to a standard open auction that treats all bids at face value, low-cost sellers would often gain less, and high-cost sellers would occasionally gain more (even though overall costs to the Government may decrease) in these alternative auctions.

EXPERIMENTS ON AUCTION PERFORMANCE USING ALTERNATIVE DESIGNS

Given the complexity of conservation programs—with their variety of potential participants that have partially observable reservation prices—theoretical models that can predict the performance of different enrollment mechanisms are often unavailable. Even when theory exists, its predictions may depend on the specific characteristics of the pool of potential participants. Thus, economic experiments are often used to investigate the performance of alternative auction mechanisms.

A number of economic experiments on conservation auctions have investigated the effectiveness of auction design. One line of inquiry focuses on selection under asymmetric information with parcel and landowner heterogeneity. Theoretical (Wu and Babcock, 1996; Smith, 1995) and experimental work (Arnold et al., 2013) indicate that heterogeneity generates a systematic tendency for landowners of lower quality parcels to make offers that yield substantial information rents (extra profits due to information on the attributes of their land known only by the landowner).

A second area of experimental investigation compares the performance of discriminatory auctions with fixed-price procurement. Generally, discriminatory auctions outperform uniform-price auctions (Cason and Gangadharan, 2004; Schilizzi and Latacz-Lohmann, 2007). There is concern, however, that this advantage dissipates over time as participants gain experience (Hailu and Schilizzi, 2004).

A third area of inquiry is the role of public and private information in conservation auctions. The depth of information about the goals and budget of the buyer, and the costs and preferences of possible participants in an auction, can affect auction outcomes in a variety of ways. The impacts of information on auction outcomes have been explored in a number of experimental studies. For example, Cason et al. (2003) find that sellers' rents increase as they gain experience and garner more information about demand and seller heterogeneity. They also find that sellers tend to inflate their requested payments more for projects with a high value to the buyer. Consequently, less information about buyer value may increase auction efficiency. Messer et al. (2013) find that, for a discriminatory reverse auction, a limited-information setting leads to greater market efficiency than either no-information or full-information settings (i.e., providing just the right amount of information can encourage market competition while limiting market manipulation).

While the existing literature on experimental studies is informative, it is far from complete. In particular, it does not address the effectiveness of enrollment mechanisms that use available information with asymmetric sellers. To examine these kinds of questions, we review findings from experimental studies conducted by ERS and academic researchers on the effectiveness of bid caps, quota auctions, and reference-price auctions in improving auction performance. The box, "Basic Factors of Experimental Design," highlights design decisions that experiments should consider.

Box: Basic Factors of Experimental Design

Economic experiments for policy and behavioral research have become popular over the last few decades. Economic experiments can provide replicability, laboratory controls, and the ability to incorporate incentives. Experiments can provide insights into policy options, as well as test and further develop theories of human economic behavior.[1]

The specific details of an experiment's design can be important. For example, in many lab experiments, individual subjects are provided with real-money payments that are dependent upon outcomes in the experiment.[2] How high should compensation be to induce realistic behavior (that is, behavior that is representative of behavior outside the lab)? How should the experiment be presented—i.e., what is the proper mix of text, visual tools, and verbal instruction? Does the setting matter—i.e., does behavior depend on where the experiment takes place, such as remotely in one's home, in a university's specialized laboratory, or in a community meeting hall?

A well-designed experiment considers these and other factors; the goal of proper experimental design is to create a cost-effective experiment that allows the experimenter to learn by evaluating hypotheses. For example, laboratory experiments tend to be inexpensive and relatively easy to design. In contrast, large-scale field experiments—experiments that take place in the context of a Government program or policy—can be very expensive. The quality of the information obtained by the two methods is not uniform, of course. While a lab experiment is capable of providing the researcher with feedback on how subjects respond immediately to small-stakes financial risk, lab experiments cannot determine the long-term effects of CRP participation on an individual's well-being.

[1] Shogren (2004, p. 1218) described the process as being "like a wind tunnel to test airplane design, lab experiments provide a test bed for what is called *economic design*—the process of constructing institutions and mechanisms to examine resource allocation."

[2] See Smith (1976) and Smith (1982) for details of using induced-value theory to conduct lab experiments.

Bid Cap Experiment

Hellerstein and Higgins (2010) considered the impact of bid caps, such as the CRP's SRR. To simulate a population with asymmetric potential program enrollees, participants in the experiment were endowed with tickets of different types.[29] Each ticket type differed in its range of costs.[30] During each round, participants could offer their tickets for sale by choosing an asking price and selecting a level of quality improvements. If accepted, the participant receives the difference between the asking price and the costs (including the costs of providing the selected level of quality).

In line with the theoretical results, these experiments found that stringent bid caps (where 20 percent of tickets have costs greater than their bid cap) significantly reduced performance as many low-cost sellers with bid caps below actual cost chose not to make offers. Performance was best when 90 percent of tickets had bid-cap costs greater than their costs (so 10 percent could not be profitably offered). When quality improvements are valued as much as cost minimization, a bid cap equal to the maximum cost of a ticket of that type could yield the most effective results; higher bid caps encourage more participants to ask for more money and increase their quality commensurately. In fact, very high bid caps (twice the ticket-type maximum) yielded performance similar to the stringent bid cap scenario.

These results suggest that using average rental rates (e.g., the CRP's SRR) as a bid cap—which implies that 50 percent of parcels will have agricultural returns that are above the bid cap—may be counterproductive. Instead, a bid cap that is set high—say, equal to an estimate of the maximum cost across all potential participants—could be better. However, alternative design mechanisms (such as quota or reference-price auctions) may function even better than a well-constructed bid cap.

Quota Auction Experiment

Quota auctions group together sellers based on their observable features, and a fraction of each group—those with the highest bids—are rejected. Higgins et al. (2014) compared discriminative auctions to a quota auction. They began by applying a theoretical model to simulate auction performance under different conditions (such as the distribution of underlying costs).[31] As expected, their results suggest that low-cost sellers extract information rents because of the observable heterogeneity of high-cost sellers. They found that, in a quota auction (with the highest bid of each type automatically rejected), low-cost sellers significantly lower their bids while high-cost sellers slightly inflate their bids.[32] This illustrates a general finding (McAfee and McMillan, 1987) that the returns to enforcing a small bit of competition, by creating artificial scarcity, are highest when the amount of competition starts out low, as it does among the low-cost sellers in auctions without a quota.[33]

Experiments confirm these results. As described in appendix 1, laboratory experiments were conducted that looked at open auctions and quota auctions. In each session, five participants were given a low-cost ticket, and five were given a high-cost ticket. Each ticket type is characterized by a cost range and each ticket's cost pulled from this range. Participants knew the cost ranges of each type, but did not know the costs of other participants' tickets. Around 50 percent of offers were accepted, with accepted offers paid their bid (e.g., the auction was discriminative).

Figs. 2 and 3 graph the relationship between bids and ticket costs across a number of experimental sessions. If sellers bid to provide environmental services at just their actual cost, then the points should fall on the 45-degree (Cost=Bid) line. Bids that are above costs (that yield information rents) lie above the 45-degree line. Under a hypothetical, perfectly cost-effective auction design, offers would be equal to cost, so all points would be clustered along a 45-degree line from the origin. The vertical distance between the 45-degree line and the actual bid is a measure of information rents accruing to ticket owners. In particular, given that only a fraction of offers are accepted, the sum value of this distance in the lowest bids is a measure of cost-effectiveness.

The results of the standard open (discriminative) auction (fig. 2) illustrate the behavior of low-cost sellers. Because low-cost sellers have a high probability of acceptance, they bid proportionally higher above their costs—the lower a seller's cost, the higher the economic profits. Due to this, the bids of low-cost sellers tend to be clustered around a horizontal line equal to the approximate maximum accepted bid.[34] However, due to the extra competition introduced by the quota (fig. 3), low-cost bids tend to be clustered closer to the 45-degree line. Low-cost sellers still bid on a horizontal line but this line has shifted down substantially in comparison to the open auction.

Fig. 4 plots estimated bidding functions using the data displayed in figs. 2 and 3, differentiated by ticket type (type A tickets have lower average cost, type B tickets have higher average cost). Significantly, type A ticket holders (low-cost sellers) submit lower bids during quota auctions than during open auctions.

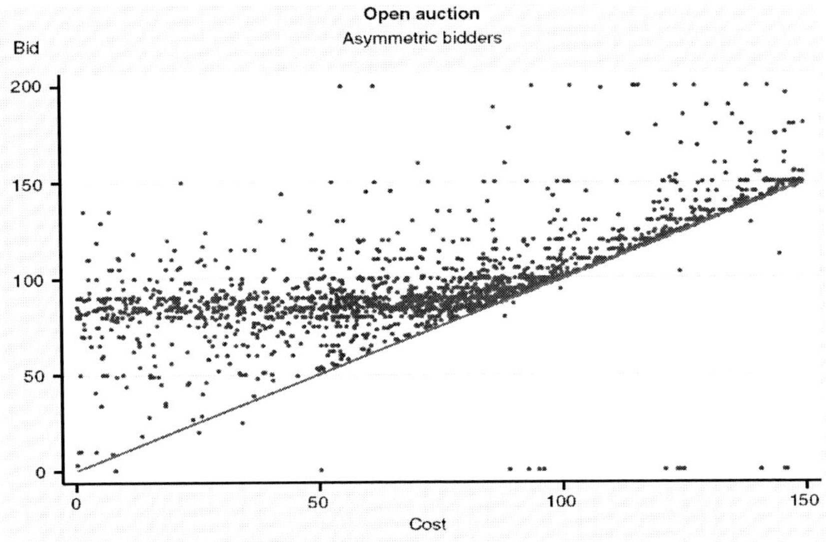

Note: 45-degree line is cost=bid.
Source: USDA, Economic Research Service.

Figure 2. Auction bids versus seller costs: Results of open auctions.

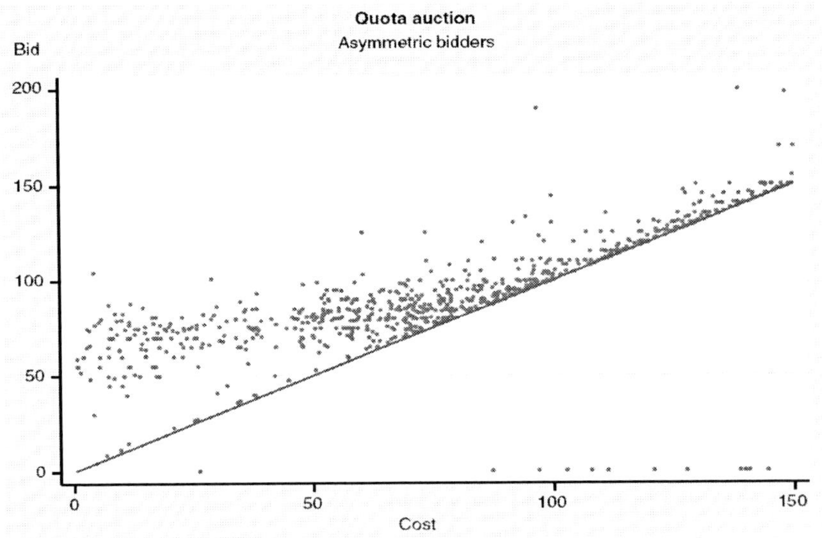

Note: 45-degree line is cost=bid.
Source: USDA, Economic Research Service.

Figure 3. Auction bids versus seller costs: Results of quota auctions.

These findings suggest that quota auctions are more cost effective than open auctions—low-cost sellers are incentivized by competitive forces to bid closer to their costs, while high-cost sellers don't change their bids by much, resulting in aggregate program savings. When estimated across all the experimental sessions, the procurement cost of quota auctions was 8.7 percent less than in the open auctions (which is greater than the 4.1-percent reduction the theoretical model predicted in appendix 1). These results suggest that, if between-group heterogeneity is sufficiently high, substantial savings can be achieved by imposing quotas. Further, if between-group heterogeneity is relatively low, total procurement cost is unlikely to increase by a substantial amount through a quota.

Reference-Price Auction Experiment

In reference-price auctions, a nonbinding estimate of value (a reference price) is assigned to all potential offers, with offers ranked based on the difference between the asking price and the reference price. The potential benefits and costs of reference-price auctions were also examined in a set of experiments (appendix 1). Compared to the quota experiments discussed above, a greater variety of types of sellers were used (low cost, high cost, and a high- and low-variance medium cost). Furthermore, participants were able to improve the quality of their offers as well as choose an asking price. This feature attempts to mimic programs like the CRP, where mere enrollment in a program is not the whole story—the benefits from conserving a given parcel can be enhanced.

The distribution of average costs and average profits within each experimental round are in fig. 5. The line graphs suggest that the reference price and quota auctions are more cost effective than the open (discriminative) auction—the average profit rates tend to be lower (lower values of the profit rate imply more cost-effective results).

A more rigorous test is provided by regression analyses that compare performance across different treatments. Several types of regressions were performed using both average profit for accepted offers and expenditure efficiency, which compares actual expenditures to a full-information theoretical minimum (that is, to a world where the purchaser could obtain all desired parcels at cost).[35]

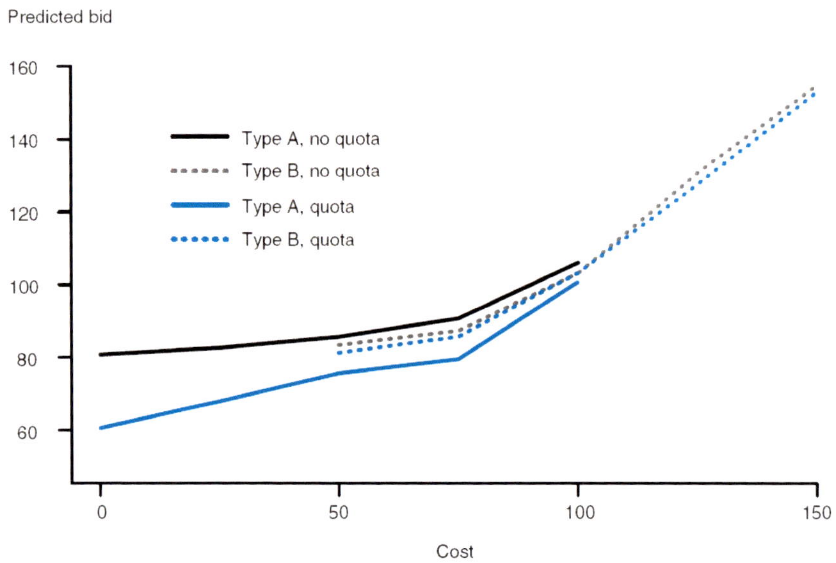

Source: USDA, Economic Research Service.

Figure 4. Bidding functions based on figs. 2 and 3.

The regressions found that cost savings from the reference price and quota auctions were statistically significant. The simplest model (a linear regression using treatment dummies) predicts that the reference-price auction has 18 percent lower (hence better) *expenditure efficiency* than an open auction, while the quota auction is 14 percent better (table A1.4). Similarly, when considering average profit per offer (which abstracts from quality improvements), the reference price is predicted to have \$10.50 less in average profit per offer than the open auction, while the quota auctions yield an average profit that is \$7.70 lower. Other models, such as panel models and difference-in-differ ence models (that compare differences between rounds to control for the unobservable attributes of participants—see appendix 1) yielded similar, though slightly smaller, results.

By way of comparison, consider the overall results if these mechanisms had been used to enroll the CRP's 19.7 million general signup acres (as of September 2014) and the per-acre cost reductions suggested by these experiments were realized. With annual rental payments of \$1.007 billion (circa September 2014), a reference-price mechanism that reduced costs by 18 percent would yield a \$181- million reduction in program expenditures, while a quota auction that reduced costs by 14 percent would yield a $141-million reduction.

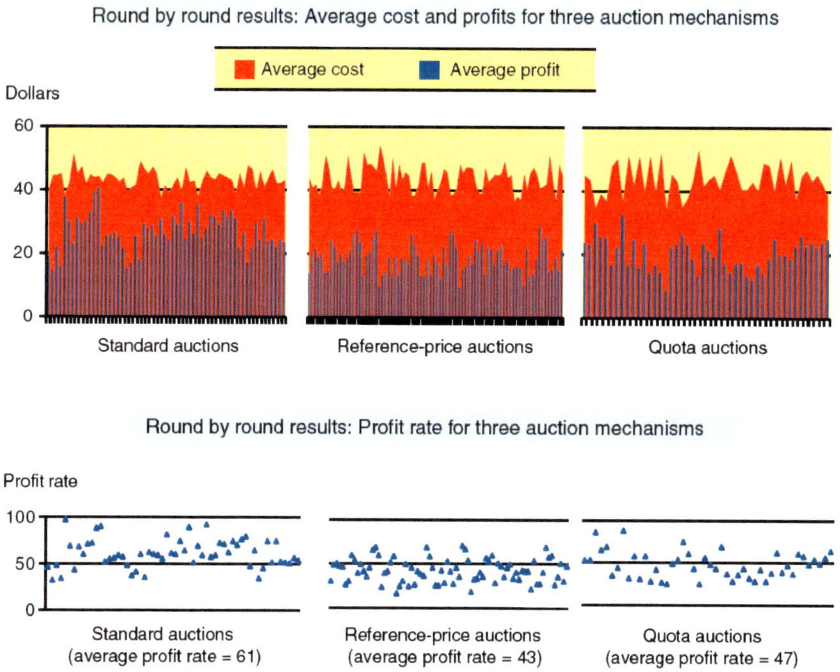

Notes: The upper three plots show the per-round average cost of accepted offers (in red) and the average profit for each accepted offer (in purple) for three kinds of auction: open (discriminative), reference price, and quota auctions. The lower three plots show the average profit rate—defined as average profit divided by the average cost. Source: USDA, Economic Research Service.

Figure 5. Average profit and cost in open, quota, and reference-price auctions.

CONCLUSION

Auctions offer an appealing means of choosing whom to enroll in conservation programs and determine what price to pay for services provided. Auctions can lower procurement costs, increase the quality of purchased goods and services, and facilitate the design and implementation of markets for ecosystem services. Results of economic experiments, which offer a useful means of testing alternative auction design, suggest that auctions (especially discriminative auctions) can be more effective in terms of reducing costs and maximizing environmental benefits than the simpler alternatives (such as using a single fixed price). However, much of what is currently considered good design in conservation program auctions may be sub-optimal.

In particular, the CRP's use of bid caps can be improved. Bid caps can prevent large profits (they can reduce information rents) to CRP participants with low-valued agricultural land that is highly erodible (e.g., land that will have a high EBI). However, an improperly set bid cap can reduce program participation, dissuading submission of offers for parcels targeted by the program (such as highly erodible parcels that have a greater agricultural value than the bid cap's assessment). Use of this auction mechanism can lead to higher costs and discourage improvements of land from parcels that are likely to be accepted at the bid cap, since costly improvements in quality cannot be compensated for by increasing the asking price.

Auction mechanisms that leverage observable information to discriminate among participants may improve the cost-effectiveness of conservation programs. Examples of these include quota auctions, which group similar participants together, and reference-price auctions, which relax the bid cap. In quota auctions, a fraction of each group is rejected in the first stage—this can induce competition between low-cost participants but requires accepting more high-cost participants. Reference-price auctions allow asking prices to be greater than the buyer's estimate of the seller's opportunity cost.[36] As a result, low-cost sellers are never fully excluded and they have an incentive to improve quality. The net effect, however, may be an increase in program costs if the increased competition is insufficient to deter all auction participants from increasing their bids.

Any auction mechanism involves tradeoffs. Economic experiments offer a means of testing the efficacy of alternative designs. A series of lab experiments, using university students, demonstrate that stringent bid caps could reduce auction performance as many low-cost sellers chose not to make offers. They also show that quota and reference-price mechanisms could improve program cost-effectiveness in scenarios where participants can choose both an asking price and make quality improvements to their offers. In particular, experiments using quota and reference prices yield improvements in quality-adjusted cost-effectiveness that range between 14 and 18 percent.

These experiments also indicate that, when considering auction design for an actual program, the degree of heterogeneity in costs and environmental benefits, the ability to accurately measure opportunity costs, and the extent to which the competitive behavior of the target population is not well represented by student participants in a laboratory all matter. Further experiments could yield different results, such as experiments that use a more representative sample (i.e., with rural landowners), that use an instrument that more closely resembles how decisions would be made, and that involve cost and benefit measures more reflective of agricultural profitability. Even after thorough laboratory testing, field tests where actual decisions are made about the resource (e.g., land retirement) in question may be a suitable (albeit more expensive) next step.

The literature review and experimental results presented in this study illustrate and highlight the potential effects of alternative auction designs rather than offer firm policy guidance. Policy guidance would require a much broader set of experiments that consider a number of variations in both design and underlying population characteristics.

REFERENCES

Armantier, O. January 2013. "A "Reference Price Auction" to Buy or Sell Different Assets Simultaneously," *Liberty Street Economics*, Federal Reserve Bank of New York. http://libertys-treeteconomics.newyorkfed.org/2013/01/a-reference-price-auction-to-buy-or-sell-different-assets-simultaneously.html#.U5G3CXJdVy5 (accessed July 2014).

Armantier, O., C. A. Holt, C. R. Plott. 2010. *A Reverse Auction for Toxic Assets*, Social Science Working Paper 1330, California Institute of Technology. http://authors.library.caltech. edu/20207/1/sswp1330.pdf (accessed July 2013).

Arnold, M., J. M. Duke, and K. D. Messer. 2013. "Adverse Selection in Reverse Auctions for Ecosystem Services," *Land Economics* 89(3):387-412.

Ashenfelter, O. 1989. "How Auctions Work for Wine and Art," *The Journal of Economic Perspectives* 3(3):23-36.

Asker, J., and E. Cantillon. 2008. "Properties of Scoring Auctions," *The RAND Journal of Economics* 39(1):69-85.

Athey, S., J. Levin, and E. Seira. 2011. "Comparing Open and Sealed Bid Auctions: Evidence from Timber Auctions," *Quarterly Journal of Economics* 126(1):207-57.

Athey, S., P. Cramton, and A. Ingraham. 2002. *Auction-Based Timber Pricing and Complementary Market Reforms in British Columbia,* MDI Report for

British Columbia Ministry of Forests. http://works.bepress.com /cramton/119 (accessed July 2014).

Ausubel, L., P. Cramton, E.-F. Ozbay, N. Higgins, E. Ozbay, and A. Stocking. 2014 (Scheduled). "Common-Value Auctions with Liquidity Needs: An Experimental Test of a Troubled Assets Reverse Auction," in *Handbook of Market Design*, Z. Neeman, A. Roth, and N. Vulkan, eds. Oxford: Oxford University Press.

Ayres, I., and P. Cramton. 1996. "Deficit Reduction Through Diversity: How Affirmative Action at the FCC Increased Auction Competition," *Stanford Law Review* 48(1996):761-815. http://works. bepress.com/cramton/75 (accessed July 2014).

Back, K., and J. F. Zender. 1993. "Auctions of Divisible Goods: On the Rationale for the Treasury Experiment," *The Review of Financial Studies* 6(4):733-64.

Bajari, P., and A. Hortacsu. 2003. *Economic Insights from Internet Auctions: A Survey*, NBER Working Paper Series 10076, National Bureau of Economic Research, Cambridge, MA.

Bajari, P., and A. Hortacsu. 2003. "The Winner's Curse, Reserve Prices, and Endogenous Entry: Empirical Insights from eBay Auctions," *The RAND Journal of Economics* 34(2):329-55.

Beshears, J., J. Choi, D. Laibson, and B. Madrian. 2013. "Simplification and Saving," *Journal of Economic Behavior and Organization* 95:130-45.

Campbell, C., and D. Levin. 2006. "When and Why Not to Auction," *Economic Theory* 27(3):583-96.

Cason, T.N., and L. Gangadharan. 2004. "Auction Design for Voluntary Conservation Programs," *American Journal of Agricultural Economics* 86(5):1211–17.

Cason, T., L. Gangadharan, and C. Duke. 2003. "A Laboratory Study of Auctions For Reducing Non-Point Source Pollution," *Journal of Environmental Economics and Management* 46(3):446–71.

Coase, R. H. 1959. "The Federal Communications Commission," *The Journal of Law & Economics* 2:1-40. http://old.ccer.edu.cn/download/7874-1.pdf (accessed July 2014).

Congdon, W., J. Kling, and S. Mullainathan. 2011. *Policy and Choice: Public Finance Through the Lens of Behavioral Economics*, Brookings Institution, Washington, D.C. http://www.brookings. edu/~/media (accessed July 2014).

Corrigan, J., A. C. Drichoutis, J. L. Lusk, R. M. Nayga, Jr., and M. C. Rousu. 2011. "Repeated Rounds with Price Feedback in Experimental Auction

Valuation: An Adversarial Collaboration," *American Journal of Agricultural Economics* 94(1):97-115.

Cramton, P. 1998. "Ascending Auctions," *European Economic Review* 42(3–5):745-56.

Cummings, R., and C. Holt. 2004. "Using Laboratory Experiments for Policymaking: An Example From The Georgia Irrigation Reduction Auction," *Journal of Policy Analysis and Management* 23(2):341-63.

Depiper, G. S., N. Higgins, D. W. Lipton, and A. Stocking. 2013. "Auction Design, Incentives, and Buying Back Maryland and Virginia Crab Licenses," *Canadian Journal of Agricultural Economics/Revue Canadienne d'Agroeconomie* 61(2):353-70.

Douglas D. D., and C. A. Holt. 1992. *Experimental Economics.* Princeton: Princeton University Press.

Eigenraam, M. 2005. *EcoTender: Paying for Ecosystem Services, not Lemons,* The Ecosystem Marketplace: A Forest Trends Initiative. http://www.ecosystemmarketplace.com/pages/dynamic/ article.page.php?pagejd=3947&section=home&eod=1 (accessed July 2014).

Eigenraam, M., C. Beverly, G. Stoneham, and J. Todd. 2005. *Auctions For Multiple Environmental Outcomes, From Desk to Field in Victoria, Australia*, Energy and Earth Resources, Department of State Development, Business and Innovation, Victoria, Australia. http://www. energyandresources.vic.gov.au/about-us/publications/ economics auctions -for-multiple-environmental-outcomes,-from-desk-to-field-in-victoria,-australia (accessed July 2014).

Ferraro, P. J. 2008. "Asymmetric Information and Contract Design for Payments for Environmental Services," *Ecological Economics* 64(4):810-21.

Filippos, E., A. M. Espin, and P. Brañas-Garza. 2013. "Experimental Subjects are Not Different," *Scientific Reports* 3(1213). http://www.nature.com /srep/2013/130214/srep01213/full/srep01213. html (accessed July 2014).

Fuerbringer, J. 1992. "Credit Markets; New Process For Auctions To Be Tested," *The New York Times*. http://www.nytimes.com/1992 /09/04 /business/credit-markets-new-process-for-auctionsto-be-tested.html?module=Search&mabReward=relbias%3Ar (accessed July 2014).

Hailu, A., and S. Schilizzi. 2004. "Are Auctions More Efficient Than Fixed Price Schemes When Offerers Learn?" *Australian Journal of Management* 29(2):147-68.

Hamilton, J. 2010. *Conserving Data in the Conservation Reserve: How A Regulatory Program Runs on Imperfect Information*, Resources for the Future, Washington, DC.

Hellerstein, D. 2012. "Conservation Reserve Program: Status and Trends," in *Agricultural Resources and Environmental Indicators, 2012,* EIB-98, U.S. Department of Agriculture, Economic Research Service. http://www.ers.usda.gov/media (accessed July 2014).

Hellerstein, D., and N. Higgins. 2010. "The Effective Use of Limited Information: Do Offer Maximums Reduce Procurement Costs in Asymmetric Auctions?" *Agricultural and Resource Economics Review* 39(2):288-304.

Herriges, J., and C. Kling. 1999. *Valuing Recreation and the Environment.* Northhampton, MA: Edward Elgar Publishing.

Herzel, L. 1951. "Public Interest and the Market in Color Television Regulation," *University of Chicago Law Review* 18(1951):802-16.

Higgins, N., M. Roberts, and D. Hellerstein. 2014. *Using Quotas to Enhance Competition in Asymmetric Auctions: A Comparison of Theoretical and Experimental Outcomes*, Working Paper No 14.9, Department of Economics, University of Hawai`i at Manoa. http://www. economics (accessed July 2014).

Horowitz, J., L. Lynch, and A. Stocking. 2009. "Competition-Based Environmental Policy: An Analysis of Farmland Preservation in Maryland," *Land Economics* 85(4):555-75.

Kagel J., and D. Levin. 1993. "Independent Private Value Auctions: Seller Behavior in First-, Second- and Third-Price Auctions with Varying Numbers of Sellers," *The Economic Journal* 103(419):868-79.

Kagel, J., and J. Richard. 2001. "Super-Experienced Bidders in First-Price Common-Value Auctions: Rules of Thumb, Nash Equilibrium Bidding, and the Winner's Curse," *The Review of Economics and Statistics* 83(3): 408-19.

Kirwan, B., R. Lubowski, and M. J. Roberts. 2005. "How Cost Effective are Land Retirement Auctions? Estimating the Difference between Payments and Willingness to Accept in the Conservation Reserve Program," *American Journal of Agricultural Economics* 87(5):1239-47.

Klemperer, P. 2002. "What Really Matters in Auction Design," *Journal of Economic Perspectives* 16(1):169-89.

Krasnokutskaya, E., and K. Seim. 2011. "Bid Preference Programs and Participation in Highway Procurement Auctions," *American Economic Review* 10(6):2653-86.

Krishna, V. 2010. *Auction Theory: Second Edition*. London: Academic Press.

Latacz-Lohmann, U., and C. van der Hamsvoort. 1997. "Auctioning Conservation Contracts: A Theoretical Analysis and an Application," *American Journal of Agricultural Economics* 79(2):407–18.

Latacz-Lohmann, U., and S. Schilizzi. 2005. *Auctions for Conservation Contracts: A Review of the Theoretical and Empirical Literature*, Report to the Scottish Executive Environment and Rural Affairs Department.

Latacz-Lohmann, U., and S. Schilizzi. 2007. "Quantifying the Benefits of Conservation Auctions," *EuroChoices* 6(3):32-9.

Lewis, G., and P. Bajari. 2011. "Procurement Contracting with Time Incentives: Theory and Evidence," *The Quarterly Journal of Economics* 126(3):1173-1211.

Madhavan, A., and V. Panchapagesan. 2000. "Price Discovery in Auction Markets: A Look Inside the Black Box," *Review of Financial Studies* 13(3):627-58.

Marion, J. 2009. "How Costly Is Affirmative Action? Government Contracting and California's Proposition 209," *Review of Economics and Statistics* 91(3):503–22.

McAfee, P. R., and J. McMillan. 1987. "Auctions and Bidding," *Journal of Economic Literature* 25(June 1987):699-738.

Menezes, F. M., and P. Monteiro. 2005. *An Introduction to Auction Theory*. Oxford: Oxford University Press.

Messer, K., J. Duke, and L. Lynch. 2013. "Applying Experiments to Land Economics: Public Information and Auction Efficiency in Ecosystem Service Markets," in *Oxford Handbook of Land Economics*, J. Dune and J Wu, eds. Oxford: Oxford University Press.

Mester, L. J. 1995. "There's More than One Way to Sell a Security: The Treasury's Auction Experiment," *Federal Reserve Bank of Philadelphia Business Review*, July/August:3-17.

Milgrom, P. 1989. "Auctions and Bidding: A Primer," *The Journal of Economic Perspectives* 3(3):3-22.

Milgrom, P. 2004. *Putting Auction Theory to Work*. Cambridge: Cambridge University Press.

Normann, H.-T., T. Requate, and I. Waichman. 2013. *Do Short-Term Laboratory Experiments Provide Valid Descriptions of Long-Term Economic Interactions? A Study of Cournot Markets*, Social Science Research Network. http://dx.doi.org/10.2139/ssrn.2057675 (accessed July 2014).

Ozbay, E. Y. 2007. "Auctions with Anticipated Regret: Theory and Experiment," *American Economic Review* 97(4):1407-18.

Paarsch, H. J., and H. Hong. 2006. *An Introduction to the Structural Econometrics of Auction Data.* Cambridge, MA: MIT Press.

Polasky, S., D.J. Lewis, A.J. Plantinga, and E. Nelson. 2014. "Implementing the Optimal Provision of Ecosystem Services," *Proceedings of the National Academy of Sciences of the United States of America* 111(17): 6248-53. http://www.ncbi.nlm.nih.gov/pmc/articles/PMC4035974/ (accessed October 2014).

Rothkopf, M. 1969. "A Model of Rational Competitive Bidding," *Management Science* 15(7):362-73.

Schilizzi, S., and U. Latacz-Lohmann. 2007. "Assessing the Performance of Conservation Auctions: An Experimental Study," *Land Economics* 83(4):497-515.

Shogren, J.F. 2004. "Incentive Mechanism Testbeds: Discussion," *American Journal of Agricultural Economics* 86(5):1218–19.

Sill, D. K. 1994. "Managing the Public Debt," *Federal Reserve Bank of Philadelphia Business Review* July/August:3-13.

Smith, R.B.W. 1995. "The Conservation Reserve Program as a Least-Cost Land Retirement Mechanism," *American Journal of Agricultural Economics* 77(1995):93-105.

Smith, V. L. 1976. "Experimental Economics: Induced Value Theory," *The American Economic Review* 66(2):274-79.

Smith, V. L. 1982. "Microeconomic Systems as an Experimental Science," *The American Economic Review* 72(5):923-55.

Sommerville, M. M., J. P.G. Jones, and E. J. Milner-Gulland. 2009. "A Revised Conceptual Framework for Payments for Environmental Services," *Ecology and Society* 14(2):34. http:// www.ecologyandsociety.org/vol14/iss2/art34/ (accessed July 2014).

Stoneham, G., V. Chaudhri, A. Ha, and L. Strappazzon. 2003. "Auctions for Conservation Contracts: An Empirical Examination of Victoria's BushTender Trial," *Australian Journal of Agricultural and Resource Economics* 47(4):477-500. http://www.ingentaconnect.com/content/bpl/ajar/2003/00000047/00000004/art00003 (accessed July 2014).

Stubbs, M. 2013. *Agricultural Conservation: A Guide to Programs*, Report R40763, U.S. Library of Congress, Congressional Research Service.

Thaler, R. 1988. "Anomalies: The Winner's Curse," *Journal of Economic Perspectives* 2(1): 191–202.

U.S. Department of Agriculture, Farm Service Agency. 2012. *CRP Fact Sheet, Dec 2012.* http:// www.fsa.usda.gov/Internet/FSA_File/dec2012onepager.pdf (accessed July 2014).

U.S. Department of Agriculture, Farm Service Agency. 2014. *Conservation Programs, Conservation Reserve Program.* http://www.fsa.usda.gov/FSA/webapp?area=home&subject=copr&topic=crp (accessed July 2014).

U.S. Department of Agriculture, Farm Service Agency. 1995-2013. *Conservation Reserve Program, Confidential Program Data.*

U.S. Government Accountability Office. 2013. *Reverse Auctions: Guidance Is Needed to Maximize Competition and Achieve Cost Savings*, GAO-14-108.

Van Noordwijk, M., B. Leimona, R. Jindal, G. B. Villamor, M. Vardhan, S. Namirembe, D. Catacutan, J. Kerr, P. A. Minang, and T. P. Tomich. 2012. "Payments for Environmental Services: Evolution Toward Efficient and Fair Incentives for Multifunctional Landscapes," *Annual Review of Environment and Resources Online* 37:389-420. http://www.annual review abs/10.1146/annurev-environ-042511-150526 (accessed July 2014).

Vickrey, W. 1961. "Counterspeculation, Auctions, and Competitive Sealed Tenders," *The Journal of Finance* 16(1):8-37.

Vickrey, W. 1962. "Auctions and Bidding Games," in *Recent Advances in Game Theory*, O. Morgenstern and A. W. Tucker, eds. Princeton, NJ: The Ivy Curtis Press.

The Wall Street Journal. September 4, 1992. "Treasury to Try 'Dutch' System at Its Auctions."

Wu, J., and B. A. Babcock. 1996. "Contract Design for the Purchase of Environmental Goods from Agriculture," *American Journal of Agricultural Economics* 78(November 1996): 935-45.

APPENDIX 1: AUCTION EXPERIMENTS

Between 2010 and 2013, USDA's Economic Research Service sponsored a set of experiments examining quota and reference-price-auction mechanisms, using students as participants. One set considered quota auctions, while a subsequent set expanded on these results by also considering reference-price auctions, a greater range of choices, and the ability to select both an asking price and a level of quality.

Each of these sets comprised a number of approximately 1-hour long sessions held in the University of Maryland-College Park's experimental economics lab. Each session typically included 10 participants (sellers) completing up to 30 rounds (each round represented a single auction). After instructions and practice rounds, the subjects participated in computerized auctions for real money. During each round—which lasted between 1 and 2 minutes—each seller received one or two tickets that they could then submit for sale to the buyer.[37] When submitting a ticket, the participant chose an asking price, and in some rounds, a number of quality points (which cost them money but made it more likely that their ticket would be accepted). Depending on the choices made by the participants, tickets were either accepted or rejected. Accepted tickets were paid an amount based on the seller's asking price, minus the cost of the ticket and any quality points purchased.

Each seller was given the same *types* of tickets in each round. Each type of ticket had a range from which a cost was randomly drawn for each round. Sellers knew the types of the tickets they received, the cost range of each type of ticket, and the total number of each type (across all other sellers). However, the actual cost of each ticket was known only by the seller submitting the ticket for sale. This structure is designed to create an asymmetric environment where there are different classes of sellers and limited public information.

Quota Auction Experiments

A quota auction imposes a limit on the number of winning offers that can come from any single category (in this case, type of ticket). In these experiments, the auctioneer specified, before each round began, that he or she will accept no more than *x* offers from each type of ticket. Tickets that failed this test—those ranked amongst the lowest of all the tickets submitted of this type (in this round)—were rejected regardless of their overall ranking. Remaining tickets were then re-ranked, with a fraction (i.e., 50 percent) selected. Note that neither of these procedures entailed rebidding. After selections are made, sellers were told whether their tickets were accepted or rejected, and why they were rejected.

Although theoretical models of symmetric auctions are well developed, models are lacking for more complex auction structures. As described in the box, "Constrained Strategic Equilibrium" (CSE), the CSE approach was used to simulate seller behavior in a quota auction. This method shows conditions under which a quota can be an effective means of reducing procurement costs in situations with high seller asymmetry, while posing little risk of increasing procurement costs when groups of sellers are similar.

Box: Constrained Strategic Equilibrium

In order to demonstrate the effect of quotas in the simplest possible environment, consider two groups of sellers that are observably different: type *A* (low cost) sellers with costs ranging between C_{A0} to C_{A1}, and type *B* (high cost) sellers whose costs range between C_{B0} to CB_1. Sellers perceive the other group as similar to themselves but with a different average cost. The buyer will accept the lowest *m* offers submitted by the *N* total sellers.

To simulate behavior under this mechanism, a markup model was used (Higgins et al., 2014).[1] The markup strategy posits that a seller with an opportunity cost c_i will submit an offer that is some constant markup of that cost, Θc_i, the parameter Θ being chosen intelligently to maximize expected profits. By restricting attention to strategies that follow a particular functional form, Higgins et al. (2014) were able to numerically estimate bidding strategies.

Equilibrium approximations in three cases of asymmetry (low, medium, and high) were calculated with five type *A* and five type *B* sellers. In all three cases, type *B* sellers request lower payments than type *A* sellers with the same cost.[2] Moreover, as asymmetry moves from low to high (as the difference in mean costs between type *A* and *B* sellers increases), type *A* sellers increase their bids in response to the weaker competition provided by type *B* sellers. The takeaway is that low-cost sellers are extracting windfall profits because of the observable heterogeneity of the high-cost sellers. The greater the observable difference between sellers, the greater the profit extracted by the low-cost sellers.

The simulated results change under a quota auction that uses a simple rule: no more than four type *A* sellers and four type *B* sellers can be accepted. The imposition of a quota has a pronounced impact on bidding behavior. When faced with both within- and between-group competition, low-cost sellers submit bids much closer to their actual costs.

The results are summarized in the following table and figure; the figure plots the bid functions before and after the imposition of a quota in the medium asymmetry case. It illustrates how type *A* sellers submit significantly lower bids in an auction with a quota than in an open auction, while type *B* sellers inflate their bids slightly to reflect their increased chances of winning under the quota regime. The table summarizes overall procurement costs, highlighting how the quota mechanism is most pronounced when asymmetry is high.

[1] The approach is known as the Constrained Strategic Equilibrium approach (Rothkopf, 1969; Kagel and Richard, 2001).

[2] This phenomenon has been described as "weakness leads to aggression" (Krishna, 2010; p. 47).

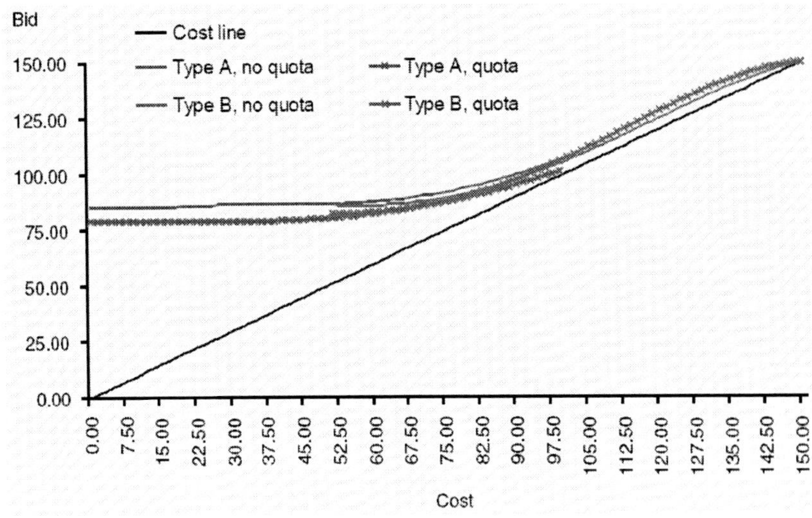

Source: USDA, Economic Research Service.

Figure. Medium asymmetry auction comparison strategies.

Table. Expected procurement costs from the simulations

	Open auction ($)		Quota auction ($)	
Symmetric auction	381.6	(17.9)	380.4	(18.2)
Low-asymmetry auction	457.3	(16.1)	452.2	(17.3)
Medium-asymmetry auction	531.5	(13.8)	509.7	(21.3)
High-asymmetry auction	643.6	(15.7)	564.8	(24.8)

Note: Standard deviations are in parenthesis.

Source: USDA, Economic Research Service.

While potentially revealing, these models depend on numeric simulations and behavioral assumptions that are difficult to statistically measure. Experiments can offer empirical validation that complements the numerical computations provided by theoretical models and simulations. In a set of 17 experimental sessions, 5 subjects were given type A tickets and 5 were given type B tickets. All subjects knew that type A sellers drew their costs randomly, with each ticket costing between $0.00 and $100.00 and each amount being equally likely to occur, while type B sellers drew costs between $50.00 and $150.00.[38]

Each session included a number of rounds. Some of the rounds used a standard auction that simply selected the lowest 6 of the 10 offers (the ticket type was ignored). Other rounds used a quota mechanism that accepted no more than four offers from each ticket type. Because every subject participated in both treatments, we can compare how the same individual performed under both types of auctions, as well as compare the performance of different subjects.[39]

The experimental results (table A1.1) confirm our computational finding that auctions with a quota are more cost effective than auctions without a quota. The difference in total procurement cost between the experimental and simulation results is more than expected for the median asymmetry case (which contains a moderate amount of systematic differences in cost). The experimental results yielded an average procurement cost that was 8.7 percent lower in auctions with a simple quota than in standard open auctions. In comparison, model-derived predictions of average procurement costs under each type of auction were higher than the experimental results, and the quota auction was only 4.1 percent lower than the standard open auction.

These results suggest several broad conclusions. First, if the between-group heterogeneity is sufficiently high, substantial savings can be achieved by imposing a quota. Second, if between-group heterogeneity is relatively low, total procurement cost is unlikely to increase by a substantial amount through the imposition of quota. Even when between-group heterogeneity is at its lowest (a symmetric auction), imposing quotas does not increase costs substantially because the expected margin for offer acceptance is relatively unaffected by the exclusion of a few sellers. These results indicate that quota auctions can be a relatively safe mechanism from the perspective of the buyer.

Table A1.1. Experimental and simulated results from the median asymmetry case

	Experimental results: Total procurement costs		Simulated results: Total procurement costs	
	Standard auction	*Quota auction*	*Standard auction*	*Quota auction*
Average cost	$518.57	$473.30	$531.49	$509.73
Standard deviation	63.24	54.19	13.8	21.3
Number of offers	200	76	10,000	10,000

Source: USDA, Economic Research Service.

Reference-Price Auction Experiments

Following the quota auction experiments, an additional 12 laboratory sessions were conducted, with much of the structure of the first set of auctions being retained. In particular, participants sold tickets that were classified by type, each experimental session consisted of a number of rounds, and treatments changed over the course of a session. The scope of the treatments was broadened, however. First, in addition to the open (standard) and quota mechanisms, a reference-price auction mechanism was considered. Second, in addition to selecting an asking price, sellers could choose to supply between 0 and 20 quality points for each offered ticket.[40] As detailed in table A1.2, four ticket types were used. Each seller received two tickets in each round, with one being a type A or type B ticket, and the other a type C or type D ticket.

When a reference-price auction mechanism was used, each ticket received a reference price that depended on the ticket type—all tickets of a given type had the same reference price. For each offered ticket, a score is computed that uses the asking price, the number of quality points added, and (in reference-price rounds) the reference price. Table A1.3 lists how scores were computed, with **lower** scores receiving a **higher** rank. At the end of an auction round, half of the offers were accepted, with accepted tickets paid the difference between their asking price and the ticket's cost (minus the cost of any quality points added). The reference price has no effect on earnings, it only affects the ranking (hence the probability of ticket acceptance).

The same sets of tickets were used in each round across the different sessions. This regularity facilitated the use of regression modeling using pairwise comparison (which compares two experiments that only differ in one attribute) since cross-session pairs can be formed that only differ in treatment (the cost and other attributes of the tickets depend on the round and not the session).

Table A1.2. Reference-price experiment, ticket details

Ticket type	Cost range ($)	Reference price ($)
A: Low cost with low variance	30- 45	39
B: Low cost with medium variance	20-65	45
C: Medium cost with medium variance	35-95	71
D: High cost with high variance	40-150	94

Notes: A ticket's actual cost is randomly drawn from the range of its type. Reference price, when used, is the same for all tickets of the same type.
Source: USDA, Economic Research Service.

Table A1.3. Reference-price experiment, treatment details

Treatment	Details	Score calculation (lower scores are better)
Quota	A two-stage acceptance procedure: 1: Within each ticket type, the highest scoring ticket is dropped 2: The survivors are pooled, and the best (lower scoring) are accepted	Score = BID - q *(q=quality points purchased)*
Reference price	A reference price is assigned to each ticket. • Bids below the reference price: score is reduced • Bids above the reference price: score is increased	Score = BID - 1.5q + (BID – referencePrice) *(q=quality points purchased) 1*

[1] The use of a multiplier of 1.5 on q adjusts for the expenditure needed to obtain q. Otherwise, an increase in the bid to offset the cost of purchasing q would negate the value of q since such an increase would increase both the Bid and the (Bid-referencePrice) terms. In a sense, this multiplier adjusts the reference price so that it includes the cost of the quality points supplied.
Source: USDA, Economic Research Service.

The net effect will be an increase in cost-effectiveness if low-cost sellers submit offers but refrain from requesting payments that are well above their opportunity costs. Conversely, if a number of high-cost sellers are accepted—due to having reference prices well above their actual costs—costeffectiveness could decrease. Two aggregate metrics were used to measure cost-effectiveness. The first is average profit per accepted ticket in a given round. The second is expenditure efficiency of a given round, defined as the actual expenditure on the quantity (A) of accepted tickets, with q total quality points, divided by what it would cost to obtain A tickets if perfect information availability allowed the buyer to offer each seller their cost and ensure provision of q quality points across these offers. Average profit is simpler but doesn't consider quality changes; expenditure efficiency recognizes the cost of these quality improvements. The box, "Estimators Used in Reference-Price Auction Experiments" contains details on the models used to regress these metrics on treatment dummies and other explanatory variables.

Table A1.4 lists linear regressions using dummies to identify treatments.[41] Both regressions (for average profit and expenditure efficiency) show that the quota and reference-price auctions improve cost-effectiveness. Table A1.5 lists difference models, using average profit as the dependent variable and using dummies to identify treatments.

Box: Estimators Used in Reference-Price Auction Experiments

The basic model used in the second-wave experiments is:

$$Y_{sr} = T_{s}\beta_{T} + X_{sr}\beta_{X} + C_{r}\beta_{c} + Z_{s}\beta_{z} + \varepsilon_{s} + \varepsilon_{r} + \varepsilon_{sr} ,$$

Where:

s,r = session, and round within session

T = vector of treatment dummies (reference price and quota), only one of which is nonzero

X = round-specific variables (maxPrior)

Z = session descriptor (exper)

C = round-specific costs (uniCost and uniRatio—these have the same (or nearly the same) values in round r, regardless of session,

eps = error components: session specific (s), round specific (r), and observation specific (rs)

Y = average profit, ratio of actual expenditures to optimal (full information) cost, or ratio of actual costs (of enrolled parcels) expenditures to optimal cost.

The difference model:

$$\Delta Y_{s,r12} = \Delta T_{s,r12}\,\beta_T + \Delta X_{s,r12}\,\beta_X + \Delta C_{s,r12}\,\beta_c + \Delta\varepsilon_{r12} + \Delta\varepsilon_{s,r12}$$

The difference model uses all pairs or rounds in the same session for which at least one of the T variables differs. Thus, it is a panel model, where each panel consists of observations within a single session. Note that the first differencing uses all pairs of rounds within a session that did not use the same treatment (it is not a simple adjacent-round first difference). Definition of first differencing for a pair of observations in panel s:

x_s,r12 = x_s,r2 – x_s,r1 (where r1 and r2 are rounds).

The eps_s session-specific error component is conditioned out by the first differencing. The Z variables are also conditioned out. Note that delta T can be negative, which means a treatment was no longer used.

The treatment dummies (ΔT) can take three values:

- -1 : the treatment was discontinued,
- 0 : the treatment was in neither element of the pair, or
- 1 : the treatment was adopted

Hence, coefficients measure what happens when a treatment is adopted (or the effect of dropping a treatment).

The difference-in-difference model:

$$\Delta^2 Y_{s12,r12} = \Delta^2 T_{s12,r12}\,\beta_T + \Delta^2 X_{s12,r12}\,\beta_X + \Delta^2 \varepsilon_{s12,r12}$$

The difference-in-difference model compares pairs of rounds between sessions S1 and S2.

Each comparison uses pairs (S2, S2) that have the same first (r1) and second round (r2).

S1 must have a round that is identical (in terms of T) to a round in S2.

S1 must have a round that is different than a round in S2. Thus, the difference within a pair is compared to a difference within another pair.

Definition of difference-in-differencing for a pair of observations spanning rounds r1 and r2, in sessions S1 and S2:

$$x\_s12r1r2 = (x\_s1,r2 - x\_s1,r1) - (x\_s2,r2 - x\_s2,r1).$$

where the r2 treatments are the same, and the r1 treatments are different. Note that the difference-in-differencing:

Controls for Z (the first difference removes session-specific variables).

Controls for changes in eps_r (the within-pair changes are the same).

Controls for changes in cost structure (since r1 and r2 have very similar cost structures across all sessions), hence C is essentially conditioned out (such as uniCost and uniRatio).

The difference-in-difference compares pairs of treatments across sessions. Each pair has the same first and second rounds. Pairs that appear in a difference-in-difference share the same treatment in the first round, but differ in the second round. Thus, the differences reflect what happens when the treatment change is different.

Table A1.4. Reference-price experiment, linear regression

	Dependent variable: Average profit per offer	Dependent variable: Expenditure efficiency
Intercept	5.7 (4.8)	1.19 (9.9)
RefPrice	-10.5 (-12.9)	-0.18 (-8.9)
Quota	-7.7 (- 8.3)	-0.14 (-5.81)
maxPrior	0.21 (7.1)	0.0054 (6.8)
uniCost	-0.33 (-6.2)	-0.018 (-13.6)
uniRatio	12.5 (3.2)	0.73 (7.3)
Exper	0.034 (.21)	-0.0041 (-0.9)
R-square [f-stat]	0.58 [< 0.001]	0.66 [< 0.001]

Notes: Numbers in parentheses are t-statistics measuring each estimator's precision; numbers in brackets are F-test probabilities measuring the probability that all coefficients are actually equal to zero (using N=206 rounds). Each observation contains the results of a round (that is, the aggregate results from all submitted and accepted offers). RefPrice and Quota are dummy variables. For expenditure efficiency, coefficients are interpreted as percent change from using this auction mechanism (compared to open auctions). MaxPrior is the max accepted score in the prior auction (which seems to have an impact on bid levels, hence acquisition costs).

UniCost is what the acquisition cost would be in a uniform price auction if the price of the least-expensive rejected offer is paid to all accepted offers. UniRatio is the ratio of uniCost to the optimal (full information) cost. Exper is a measure of prior experience (number of economic classes taken and number of experiments enrolled in).

Source: USDA, Economic Research Service.

Table A1.5. Reference price experiment, difference models

	Difference	Difference-in-difference
Intercept	1.43 (109)	1.33 (7.8)
RefPrice	-9.7 (-37.2)	-6.9 (-10.1)
Quota	-7.08 (-23.0)	-5.5 (-8.1)
maxPrior	0.093 (8.1)	0.02 (1.9)
uniCost	-0.39 (-25.7)	
uniRatio	15.8 (14.1)	
Exper	3.3 (3.1)	-0.027 (-0.03)
R-square (f-stat)	0.54 (<0.001)	0.05 (<0.001)
N (number of pairs of rounds)	1822	2565

Notes: Average profit is the dependent variable in these regressions. Difference models compare difference between rounds in the same session to control for unobservable attributes of participants. Difference-in-difference compares differences between pairs of rounds across sessions to control both for the unobservable attributes of participants and variations in ticket costs. The RefPrice and Quota are dummy variables. MaxPrior is the max accepted score in the prior auction (which seems to have an impact on bid levels, hence acquisition costs). UniCost is what the acquisition cost would be in a uniform price auction if the price of the least-expensive rejected offer is paid to all accepted offers. UniRatio is the ratio of uniCost to the optimal (full information) cost. Exper is a measure of prior experience (number of economic classes taken and number of experiments enrolled in). Numbers in parentheses are t-statistics; the R-square is an overall goodness of fit, and the f-stat (in parenthesis) measures joint significance of all coefficients.

Source: USDA, Economic Research Service.

Appendix 2: Effects of the CRP's Bid Cap

This study examined Environmental Benefit Index (EBI) scores of actual offers to determine if (1) landowners with low reservation prices but relatively high EBI scores increased their bid to the bid cap when they believed their offer would still be accepted, and (2) were less likely to offer additional conservation practices as part of their bid. The EBI score can be divided into three components:

- Exogenous EBI points–strictly due to the fixed features of a parcel, such as the parcel's soil erodibility.
- Endogenous EBI points—based on the choice of conservation practices. They represent a level of environmental impacts that the landowner can choose.
- Cost points—affected by the bid. The bid cap means there are two kinds of cost points—those the landowners can control, and those they will *always receive*. Since the landowners cannot ask for more than the bid cap, they will always receive some cost points. By decreasing their bids to below the bid cap, landowners can receive more cost points.[42]

Combining the exogenous EBI points and the cost points that a landowner always receives yields what we call *automatic* points—points that a landowner has no control over and receives no matter the size of their bid or type of conservation practices installed.

To test whether landowners with a high probability of acceptance into the CRP tend to bid the maximum allowed amount, we examine the "underbid" as a function of the automatic points. Data on offers to CRP general signup 15 (in 1997) through signup 45 (in 2013) are used. The underbid is defined as the difference between the parcel's bid cap and the parcel's bid, as a fraction of the parcel's bid cap. Thus, a value of 0 means "the bid equaled the bid cap," while 0.5 means "the bid is half of the bid cap." Since the EBI's scoring formula has changed over time, the independent variable (automatic points) was normalized; it is expressed as a fraction of the maximum possible automatic points in the signup. Table A2.1 displays regression results (which are also presented visually in the figure in the box, "Drawbacks of the Bid Cap").

The regression shows a highly significant negative relationship between automatic points and underbidding. A 1-percent increase in automatic points will (on average) reduce a parcel's underbid by about 0.12 percent. When offers are ranked by the number of automatic points they received, both the number of parcels with underbids and the average size of underbids decreased as the number of automatic points increased.

To test whether landowners with a high probability of acceptance are less likely to improve the environmental quality of their offers, we examine whether endogenous points decline as automatic points increase. As noted, both the dependent (endogenous points) and the independent variable (automatic points) are normalized; they are expressed as a fraction of the maximum possible for that signup.

Regression analysis (table A2.2) shows a highly significant negative relationship between automatic points and endogenous points—a 1-percent increase in automatic points will (on average) lead to a 0.297-percent reduction in a parcel's endogenous points. This supports the conjecture that the more likely an offer is to be accepted, the less likely it is that the landowner will attempt to install more environmentally beneficial practices. An auction mechanism that did not depend on bid caps (e.g., one that allowed a landowner to increase their bid to cover the cost of a more beneficial conservation practice) could counteract this effect.

Table A2.1. Underbid fraction as function of automatic EBI

Constant	0.113 (288)
Automatic points (normalized)	-0.122 (-156)
R-square	0.029
F-stat	24,241 (prob <0.0001)
N (number of offers)	809,040

Notes: EBI refers to the Environmental Benefits Index, F-stat refers to joint significance of all parameters, and R-square is a goodness-of-fit measure. T-stat measures of coefficient precision are in parentheses. When each signup is separately regressed, qualitatively similar results are obtained (with the coefficient on automatic points ranging from -0.06 to -0.34).

Source: USDA, Economic Research Service using confidential program data from USDA, Farm Service Agency, Conservation Reserve Program, 1995-2013.

Table A2.2. Endogenous EBI points as function of automatic EBI points

Constant	0.457 (511)
Automatic points (normalized)	-0.297 (-166)
R-square	0.033
F-stat	27,481 (prob <0.0001)
N (number of offers)	799,712

Notes: EBI refers to the Environmental Benefits Index, F-stat refers to joint significance of all parameters, and R-square is a goodness-of-fit measure. T-stat measures of coefficient precision are in parentheses. When each signup is separately regressed, qualitatively similar results are obtained (with the coefficient on automatic points ranging from -0.09 to -0.51).

Source: USDA, Economic Research Service using confidential program data from USDA, Farm Service Agency, Conservation Reserve Program, 1995-2013.

End Notes

[1] USDA fiscal year budgets can be found at http://www.usda.gov/wps/portal/usda/usdahome?navid=BUDGET.

[2] Trends in USDA conservation program expenditures can be found at www.ers.usda.gov/topics/natural-resourcesenvironment/conservation-programs/background.aspx.

[3] Future steps could use different parameterizations and types of participants, a limited field test of the best-performing alternative method or methods, and a full-scale trial of the most favored alternative.

[4] Although this report considers auction mechanisms, other approaches are possible. As summarized by Ferraro (2008), these include acquiring information on observable landowner characteristics and the use of screening contracts. There is evidence (Arnold et al., 2013) suggesting that, for many conservation settings, screening contracts may perform better than auctions.

[5] See Vickrey (1961 and 1962) for the original theory of efficient auctions. See Cramton (1998) for a discussion of the relative efficiency of alternative auction designs (in this case, sealed bid versus dynamic auctions). See Madhavan and Panchapagesan (2000) for a discussion of the use of auctions specifically for the purposes of price discovery (in the context of large financial markets).

[6] See Back and Zender (1993) for a review of the literature on Treasury auctions. Accounts of the pilot's announcement can be found in the *Wall Street Journal* (1992) and Fuerbringer (1992). Contemporary accounts can be found in Mester (1995) or Sill (1994).

[7] See Paarsch and Hong (2006) for a textbook review of auction econometrics; Milgrom (1989) for a primer on auctions, including early experimental literature and formative observational work; Milgrom (2004) for an example of the practical implementation of many high-value Government auctions; and Athey et al. (2011) for a review of the auction-based timber pricing mechanism of the USDA Forest Service.

[8] Direct targeting of individual parcels can best be handled through negotiations since individual attention must be paid to each contract. Fixed-price offers are like standard markets, where the buyer announces a price and sellers can choose to sell, or not sell, at that price.

[9] Auctions of this type are sometimes referred to as scoring auctions. See Asker and Cantillon (2008) for a discussion of these types of auctions. The Conservation Reserve Program (CRP) is, effectively, a scoring auction where parts of the Environmental Benefits Index represent the quality measure. Because the CRP is a large auction and many parcels can be scored using a single technology, the costs of scoring are spread over a large number of parcels, reducing the transaction costs often observed in traditional scoring auctions.

[10] This phenomenon is known as "crowding out" (Sommerville et al., 2009). For example, landowners motivated by a stewardship ethic may voluntarily adopt environmentally friendly management methods even though these reduce their profits (van Noordjwick et al., 2012). Introduction of a market that pays landowners to adopt this management method may discourage this voluntary activity.

[11] This simple example assumes that the same amount of erosion occurs on any parcel regardless of its location. It also assumes that a monetized value—one that should capture all the

environmental benefits of reduced soil erosion—is available and does not change with the location of the parcel or as more acres are retired. In practice, this is rarely the case.

[12] See Klemperer (2002) for a discussion of auction design and Depiper et al. (2013) for Government-auction design. Latacz-Lohmann and van der Hamsvoort (1997) and Latacz-Lohmann and Schilizzi (2005) provide a discussion of theoretical and practical implementations of conservation auctions.

[13] In a uniform-price auction, all accepted offers are paid the same price (often, the price of the highest accepted offer). In a discriminatory auction, each accepted offer is paid its asking price, regardless of what other accepted offers receive (see glossary).

[14] The Federal Aviation Administration auctions did not take place as scheduled, and have not taken place as of July 2014. The mock auctions did, however, take place.

[15] The value of an oil lease, for example, is largely common—regardless of who wins the auction (and thus wins the right to drill for oil in a particular spot), the amount of accessible oil is unchanged. Since the amount of this oil is unknown before the auction (the amount of oil cannot be known for sure until a well is drilled and the oil is extracted), oil companies must estimate the amount of oil under the lease. The buyer most likely to win the auction is the buyer who has assessed the value as being highest (i.e., the winning buyer is also the buyer most likely to have been overly optimistic about the value of the lease).

[16] Other successful conservation auctions include Australia's EcoTender (Stoneham et al., 2003) and Auction for Landscape Recovery (Eigenraam, 2005), as well as Scotland's Challenge Fund Scheme (Messer et al., 2013).

[17] As of November 2014, the Conservation Reserve Program (CRP) covers about 24.2 million acres, at a cost of about $1.8 billion per year. Historical trends in the program are detailed in Hellerstein (2012) and current CRP statistics are available at the USDA Farm Service Agency's website: http://www.fsa.usda.gov/FSA/webapp?area=home&subject=copr&topic=rns.

[18] The Conservation Reserve Program bid cap is based on the Soil Rental Rate (SRR) of the parcels comprising an offer. The SRR is an estimate of the agricultural value of a parcel and is based on the county average rental rate for an acre of unirrigated cropland, adjusted to reflect the parcel's relative soil productivity. Given the limited accuracy of these component measures, the SRR can be thought of as a relatively inexpensive, but imprecise, appraisal of the opportunity cost of retiring a parcel of land from agricultural production.

[19] The Environmental Benefits Index (EBI) considers several environmental criteria, including soil erodibility, impacts on water quality, impacts on wildlife habitat, and impacts on air quality. The Conservation Reserve Program's use of an EBI reflects a balance between minimizing cost and increasing a variety of environmental services (such as reducing erosion and improving wildlife habitat).

[20] When a Conservation Reserve Program (CRP) contract expires, the landowner may re-offer the land (say, in the next general signup). Thus, land can be in the CRP for longer than 10-15 years.

[21] General signups happen nearly every year. Parcels offered but not accepted in one year can be resubmitted in other years, with the same or different bids. Thus, the impact of the single general signup is ameliorated by the possibility of trying again in another year. Continuous signups, which are noncompetitive and target particularly valuable practices or resource concerns on relatively small parcels, are also used to enroll about 20 percent of Conservation Reserve Program acreage.

[22] In bid-preference auctions, accepted offers are paid their bid. However, when determining who to accept, certain *qualified* sellers are preferred. For example, a qualified seller (such as

sellers from a designated locale) might be chosen if his or her bid is within 5 percent of the lowest unqualified bid (Krasnokutskaya and Seim, 2011; Marion, 2009). Thus, all things being equal, offers from a qualified seller are more likely to be accepted.

[23] A reference-price auction was selected by the U.S. Treasury to purchase toxic assets under the Troubled Asset Relief Program (TARP) legislation during the 2008 financial crisis. Reference-price auctions have been the subject of substantial theoretical and experimental work; see Ausubel et al. (2014) or Armantier et al. (2010). Armantier (2013) also presents a succinct summary of reference-price auctions.

[24] The Maryland Agricultural Land Preservation Foundation uses what amounts to a reference price in its procurement system. An independent audit is conducted for each parcel under consideration, a valuation estimate is produced, and all bids are normalized by their valuation estimate (Horowitz et al., 2009).

[25] What we refer to as a quota is similar to what some in the auction literature call set-asides. Often used in Government procurement (Ayres and Cramton, 1996) for both purchasing and selling multiple units, set-asides reserve some quantity to be won by qualified sellers. Qualified sellers are selected based on observable characteristics, often race or business size. While set-asides are often meant to promote social goals by encouraging participation by a minority class of sellers, they may also increase competition in an auction and enhance the outcome from the buyer's perspective (Milgrom, 2004).

[26] Well-constructed groups will minimize within-group variation of opportunity costs.

[27] In particular, when there is no quota, landowners in a low-cost group are likely to set their bid close to their estimate of the highest accepted bid. This means that they choose bids that are competitive with landowners in higher cost groups.

[28] In the case of the Conservation Reserve Program, this would mean reducing the total cost of retiring a given number of acres without reducing environmental benefits.

[29] This was a context-free experiment using induced values rather than one where endogenous (or home-grown) values occur. Context free means that neutral language was used, with no mention of agriculture or environmental quality. An equivalent experiment using landowners could introduce context by asking participants to consider "offering cropland parcels, with these characteristics, to a land-retirement program" rather than consider "offering your tickets, with this cost, to a computer."

[30] For example, the ticket cost of a participant with a type A ticket would be between $10 and $30, while the cost of someone with a type B ticket would be between $25 and $50.

[31] Higgins et al. (2014) used the Constrained Strategic Equilibrium method to closely approximate equilibrium bidding behavior. In general, theory-based predictions of auction outcomes become intractable as auction complexity increases (which occurs when participants have asymmetric costs). Hence, numerical solutions are of interest.

[32] To be credible, a quota auction should accept some high-cost sellers even if their bids are greater than a rejected low-cost seller's bid. Since this increases the odds of acceptance of high-cost sellers, high-cost sellers may submit somewhat higher bids as a side effect.

[33] Technically, the returns-to-competition function is often concave—e.g., adding an additional seller is more valuable when there are 3 sellers than when there are 300.

[34] This clustering is likely due to the multiple-round nature of the experiment. Participants could use the publically reported cutoff (highest accepted) bid in prior rounds to predict the likely cutoff bid in the current round, and set their bids accordingly (i.e., near this prediction).

[35] Expenditure efficiency is the ratio of actual expenditures over full-information expenditures. Full-information expenditures are what would be paid if the purchaser offered an amount exactly equal to the seller's cost, and all participants were willing to make offers at this

amount. Thus, values close to 1.0 signal a more cost-effective auction. Average profit is the average profit made on each accepted offer, so lower values imply better cost-effectiveness. Note that expenditure efficiency incorporates the ability of participants to change the quality of their offers.

[36] In a sense, reference prices are weak bid caps—the seller can ask for more but will be penalized.

[37] Each participant was told that the buyer was a computer and these experiments were context free—no mention was made of conservation goals or agricultural income. This approach was taken to minimize the risk that students would submit bids that expressed approval (or disapproval) of conservation as a societal goal.

[38] Costs and bids were denominated in e-dollars. Actual cash payments were based on $0.05 per e-dollar. Participants earned between $15 and $75 (depending on their bidding skills and on the costs of their tickets) for participating in the 1-hour session.

[39] We varied the order of treatments to control for learning effects, and in some sessions employed an "A-B-A"-type design to determine if individual bidding behavior within a treatment varied with experience. Each treatment ran for at least 10 rounds, allowing for learning in earlier rounds.

[40] Quality points were structured to resemble the cost-share improvements often found in conservation programs. Each quality point cost $0.50, and a maximum of 20 points could be obtained for each ticket. Since obtaining one quality point has the same effect on an offer's score as reducing the asking price by a $1.00 in these experiments, profit-maximizing participants should obtain the maximum number of quality points. Nonetheless, many participants under-purchased quality points. Understanding why this occurs, and how changing auction design might improve results for both buyers and sellers, is a topic for future research. We do note that introducing quality improvements does weaken the ability to focus on a pure difference between auction mechanisms.

[41] It appears that higher costs (larger uniCost) reduce acquisition costs (bids may not adjust quickly, so higher costs mean less profit), while higher variation (larger uniRatio) increases acquisition costs (perhaps because larger profits are possible when there are more low-cost tickets).

[42] The Environmental Benefits Index (EBI) computes these cost points using the following equation: cost points= a * (1 – (offer/b)). Both the "a" and "b" parameters are announced after all general signup offers have been received. Thus, landowners do not know precisely what their cost points will be. However, these parameters have been fairly constant over time so landowners should be able to reasonably approximate the cost points they will always receive and those they can control.

INDEX

C

D

E

F

G

H

I

L

M

N

O

P

Q

R

S

T